T0133351

The Use of Geoinformatics in Investigating the Impact of Agricultural Activities between 1990 and 2010 on Land Degradation in NE of Jordan

Chair of Remote Sensing and Landscape Information System - Faculty of Environment and Natural Resources
Albert-Ludwigs-Universität Freiburg im Breisgau, Germany

By

Majed Mahmoud Faris Ibrahim

Freiburg im Breisgau, Germany

2014

Dean and Supervisor:	Prof. Dr. Barbara Koch
Co-supervisor:	Prof. Dr. Freidreike Lang
Reviewer:	Prof Dr. Kelimeztik
Disputation:	4 June 2014

Bibliographic information published by the Deutsche Nationalbibliothek

The Deutsche Nationalbibliothek lists this publication in the Deutsche Nationalbibliografie; detailed bibliographic data are available in the Internet at http://dnb.d-nb.de .

ISBN 978-3-8325-3773-9

Logos Verlag Berlin GmbH
Comeniushof, Gubener Str. 47,
10243 Berlin
Tel.: +49 (0)30 42 85 10 90
Fax: +49 (0)30 42 85 10 92
INTERNET: http://www.logos-verlag.de

Acknowledgements

First of all, I thank Allah for enlighten my way and helping me finish this research project.

I am greatly indebted to Prof. Dr. B. Koch for accepting me as a PhD student at the department of Remote Sensing and Landscape Information Systems (FELIS), University of Freiburg and for her guide; encouragement and support throughout this study.

I also want to thank Prof. Dr. F. Lang for agreeing to act as a Co-supervision, contributions, and support during my studies and work at more analysis of soil that enhanced my capability to come up with a result oriented research.

Sincerely, I thank My brother Dr. Mohammad Ibrahim at Chemistry Department and Prof. Rida Al-Adamat at the Institute for Earth and Environmental Sciences - Al-al Bayt University- Jordan, for their valuable and critical comments that helped me in fieldwork in Jordan, I also want to express my thanks to Dr. Pawan Datta, at FELIS, for their readiness to help and fruitful discussions which contributed to the success of this work, and I remain grateful to Dr. Ahmed Yousef.

I am also grateful to Dr. Mathias Dees and Dr. Claus Peter Gross at FELIS for their valuable comments during my stud, and most importantly, I remain thank to Mr. Markus Quinten (Network administrator) at FELIS for their always charming and helpful hand.

I thank all the staff at Department of Remote Sensing and Landscape Information Systems (FELIS), University of Freiburg and Institute for Earth and Environmental Sciences, Al-al Bayt University- Jordan. I feel privileged to be the recipient of all the many new things that I have learnt and for the wonderful working atmosphere, which was positively reflected in the smooth running of this work.

During the field work in Jordan, many institutions and people have supported me. I extend my sincere thanks to them. I am much indebted to the Institute for Earth and Environmental Sciences (IEES) staff for providing me all kinds of support required during the field work. Also, special gratitude goes to the staff of the laboratory of the Soil and staff of the laboratories of the Water, Environment and Arid Regions Research Centre (WEARC) to analysing the water and soil samples in their timely, which were an important input data of my research work.

Lastly, I am grateful that the Al-al Bayt University- Jordan gave me the opportunity to pursue the PhD study in Germany through the scholarship program.

I would deeply like to express my respect and appreciation to all brothers and sister for their inspiration, encouragement and support during the execution of this work. I acknowledge a strong sense of appreciation to the souls of my parents for their patience and encouragement during their lives.

Majed Ibrahim

1 November 2013

Dedicated to:

My Parents

My Brothers and Sisters

Table of Content

List of Figures

List of Tables

Summary

The land degradation problem is one global problem that faces Jordan that has become take a large space in recent years, especially agricultural land degradation. This may worsen in the coming years. Where the agricultural sector considered is one most important of the sources of gross domestic product at the state level the only source of per capita income for people in some areas. Some agricultural areas suffer from an agricultural pressure beyond the capacity of agricultural land.

At the same time there is absolute water scarcity problem faces the Jordan so that makes them unable to meet the needs of the water. In many parts of the country the groundwater constitutes the main supplier of water and also the only source in other regions. Most of the groundwater basins in Jordan are already exploited beyond their estimated safe yield.

Many of natural factors such as (climate like water and wind) and human factors (such as the wrong agricultural practices and lack of awareness of the agricultural as well as the misuse of groundwater resources) have contributed to the degradation of soil quality and deterioration of groundwater quality in Jordan. The types of degradation represented that affect lands and groundwater (a) wrong farming practices through use and overuse of biocides and fertilizers and irrigation return flows as well as Lack of agricultural management in agricultural soil protection and conservation methods. (b) Over-pumping of water that leads to salinization.

In this research, the groundwater resources in the aquifer of study area (Yarmouk basin) have been evaluated through the use of GIS and remote sensing. Remote sensing was used to map the agricultural activities in the study area between 1990 and 2010 using several satellite images. It was estimated that the Tree and vegetation areas changed from 163 Km^2 in 1990 to 110 Km^2 in 2010 and the bare soil areas increased according on the cultivated areas from 93 Km^2 in 1990 to 146 Km^2.

The DRASTIC model was implemented within a GIS environment to investigate the groundwater vulnerability. It was found that less than 4% of the study area is located in moderate vulnerability and it could be possible source of pollution through (Farms and urbanization), and it was reached the percentage of study area that has a moderate vulnerability with no possible source of pollution is 37.6 %, on the other hand, more the 11% of the study area has high with possible sources of pollution in the same area. While reached percentage of study area that has high vulnerability with no possible source of pollution is 18.3 %. Around 29.3 % of the study area has low vulnerability with no possible source of pollution. And that one of the sources of pollution that has been studied is the salinity according to the values of the absorption of sodium, and it was found that all wells inside groundwater vulnerability zones have SAR concentrations between moderate and very high values.

In this study was created model helping in build a map of land degradation which predicts degree of degradation and the impact of wrong agricultural practices as also impact of natural factors in lands degradation, it was named (Land Degradation Degree Model or "Index"). The LDD model was implemented within a geoinformatics environment that has ssignificant correlation with P–Values 6.78e-07 and 4.454e-07. It was found that the most percentage of the area of the study area is moderate which reached 45%, the high degradation is 34% and low areas accounted for degradation 21% of the study area. And it was found that the total area of hot-spot of whole area of region reached about 9.7 km^2 that reflected high risk in study area through the results of vulnerability of groundwater and land degradation.

Several management scenarios were implemented within a GIS environment to verify the impact of agricultural practices and as well as questionnaires and interviews. Where was found that each farmer has a special policy in sustainability and reclamation as well as what is going through the region of the climatic conditions. And there are suggestions that will serve the interests of agricultural land adjust agricultural management and awareness of workers in farms and water pump control to reduce the salinity in the region. And reliance on bio-fertilizers to increase soil fertility and dimension may be possible for inorganic fertilizers, as well as Management of groundwater protection and the search for alternatives from which to relieve the pressure on use of groundwater.

Abbreviations

ºC	Celsius
Cm	Centimetre
DEM	Digital Elevation Model
dSm^{-1}	deciSiemens/meter
ECe	Electrical Conductivity extrication
ECw	Electrical Conductivity in water
ESRI	Environmental Systems Research Institute
FAO	Food and Agriculture Organization in the United Nations
GDP	Gross Domestic Product
GIS	Geographic Information System
JMD	Jordan Metrological Department
JTM	Jordan Transverse Mercator
IR	Infrared
LDD	Land Degradation Degree
MCM	Million Cubic Meters
meq/l	milliequivalents per litre
MOA	Ministry of agriculture
MWI	Ministry of Water and Irrigation
NRA	Natural Resources Authority of Jordan
RGC	Royal Geographic Center
SAR	Sodium Absorption Ratio
TDS	Total Dissolved Solids
TM	Thematic Maper

Chapter 1

Introduction

This introductory chapter is divided into four parts, the first focuses on agriculture and is divided into six sections: a review of global agriculture, agriculture in arid and semi-arid areas, agriculture in Jordan and ,the impact of human activities on agriculture and intensive use of agrochemicals, general environmental pollution in the study area and new agricultural managements technology. The second part of the introduction focuses on water issues in Jordan, and demand, the third presents the objectives of the research work and the fourth and final part axis describes the structure of this thesis.

1.1. Global Agriculture

Recent years have witnessed a growing interest in the agricultural production in developing countries, and structural changes in the flow of agricultural trade. (Aksoy and Beghin, 2005). Agriculture in the 21st century faces several challenges; prime among them the need to produce more food and fiber to cover the dietary needs of a growing population coupled with a decrease in the rural labour force. (FAO, 2009). Observations suggest that population growth in the next 25 years, estimated at about two billion people, will occur in developing countries with far reaching implications in terms of urbanization and the effects on food production systems. (McCalla, 2001). Seventy percent of the global poor are rural dwellers, a considerable percentage of whom still depend on agriculture for their income.

Production in developing countries tends to be low, carried out by small farmers using inefficient production systems compared to agriculture in temperate regions. Accordingly, there is a pressing need to introduce modern farming systems that increase productivity in a fashion that does not damage natural resources or negatively impact on the environment (McCalla, 2001). Agriculture can be a major source of growth for national economies. In agriculture-dependent countries it contributes twenty nine percent of the gross domestic product (GDP) compared to 30 percent in industrialized countries and employs around sixty five percent of the labour force (World Bank, 2008)

Broadly, investment in agriculture aims to increase the productivity and income of farmers. Public investment by governments and development partners in the agricultural sectors seeks to achieve : economic growth and poverty Farmers are the largest investors in agriculture and in the absence of good governance, appropriate incentives and essential public goods they do not invest enough, which makes agricultural investment and productivity risky (McCalla, 2001., FAO, 2012).

Global agriculture faces the challenge of identifying strategies that focus on the development of technologies, institutions and policies that promote agricultural production to its full potential an engine of growth besides monitoring changes that happen in all stages of agriculture, and studying impediments to effective production.

Considering that sustainable agriculture, forestry, and other land-use practices are essential components in sustainable development and the ever increasing requirement for greater food and energy production in the face of population growth several researchers have taken up the task of studying agricultural production in arid and semi-arid areas of the world (Jaradat, 1991; Ray, 1995 and Dietz *et al.*, 1998). The subject of this thesis is the study of agricultural production and land degradation in arid and semi-arid areas of Jordan.

1.2. Agriculture in arid and semi-arid areas

Agriculture in arid and semi-arid areas is constrained by low precipitation, high temperatures, poor or low nutrient soils and high transpiration and evaporation rates in vegetated areas. Agricultural production is often accompanied by imbalance in the natural ecosystems which renders them extremely vulnerable to inappropriate land use and exploitation, particularly where farmers resort to over-cultivate available areas of fertile land in an attempt to increase production thereby exposing these lands to the risk of climatic variations (SFSA, 2013).

Agriculture is one of the most important natural resources in arid regions of the world especially for food security. Despite the acute environmental situation and characteristics in the arid and semi-arid lands the agriculture remains the major source of livelihood in many countries. Arid and semi-arid ecosystems span over 40% of the earth's total land surface, predominantly in Africa (nearly 13 million km^2) and Asia (11 million km^2), with a continuous increase due to desertification processes, induced mainly by anthropogenic activities and/or climatic change (IPCC, 2008). (White *et al.* 2002). Three quarters of global food production occurs in dry land primarily rice, wheat, maize, sorghum, millets and potato (FAO 1999), Therefore increasing productivity in arid and semi-arid zones is vital to ensure global food security (MPRA, 2006).

However, dry land agriculture faces considerable challenges: water scarcity, land degradation, persistent poverty, climate change, lack of sufficient awareness in agriculture, among others. The main question remains the development of dry land in a manner that enables local populations to improve their livelihoods by using their resources more productively (MPRA, 2006). This can be achieved through provision of access to information, experience, appropriate technology and practices within a rewarding economic and institutional environment that encourages small–scale farmers to pay attention to the long-run stewardship of the natural resources they manage. Meeting this challenge will require farmers to have access to both domestic and international markets.

Water is increasingly becoming scarce. For instance, in South Asia international water conflicts and riots over water have become commonplace (MPRA, 2006), even at the level of farmers who are digging deeper and deeper to steal water from the wells of their neighbours. As a result, there is a lowering of water tables and a reduction in yield from wells. Semiarid areas have at least one entirely rainless month/year and the amount of rainfall ranges from 500 to1000 mm per annum in most areas. This means that conditions of water deficit, water stress, or drought are common (SFSA, 2013).

Given this scenario, water-constrained dry land agriculture critically requires the following water related interventions: a) Groundwater management and monitoring, b) adopting an efficient watershed management approach c) recharging depleted groundwater aquifers and strictly regulating groundwater extraction. d) Enlisting government support for water-saving... etc. Therefore, it is important to know the agricultural situation in the study area and in the Jordan as generally

1.3. Agriculture in Jordan

1.3.1. Background

Agriculture is considered a basic pillar of economic and social development in all countries. Agriculture has started to play a major role in the protection of the environment through the past three decades, as well as protection of bio-diversity and ensuring an environmental balance that helps in securing sustainable resource use and their preservation for future generation (UN, 2008).

Like in other countries, agriculture is a driver of development in Jordan. Due to its scarce supply of water, Jordan must pay particular attention the agriculture sector and develop its economic, social and environments aspects. Progess in the agricultural sector in

Jordan requires a multifaceted approach a view to development that goes beyond immediate economic return to consider the social and environmental benefits of the sector and its importance in term of national security, environmental managements and public health (Hjort *et al.*, 1998).

Reach land in the Jordan extends to an area of around 89.3 thousand km^2 including dead Sea and the eastern Badia, covering about 88% of the total area of the country. , Ground elevation ranges from 600 to 900 m above sea level, and the temperature varies sharply between day and night, summer and winter. Rainfall is occasionally limited to 150 mm. (MoA, 2011). Average farm sizes range from to 3.5 ha to up to 34 ha in the highlands.

1.3.2. Agriculture Policy

Agricultural policy in Jordan has undergone a number of changes in past two decades. The government put forward a policy document for the sector in 1995 that featured the following main points:

1. Management of agricultural resources, especially soil, water and vegetation in a manner that ensures protection and development, and the sustainability of production in the long term.
2. Development and optimal utiliization of natural and agricultural resources..
3. Continuous updating of legislation related to the agricultural sector if it contradicts the changing needs of agricultural development.
4. Activation and amendment of legislations governing the sector and the protection of resources (land use system, the law regulating the towns and villages, the provisions of the law on the protection of agriculture resources, recruitment of expatriate expertise).

1.3.3. Farming Problems

Problems of the agricultural sector are largely identical between year and year. The problems are similar and repeated and can be summarized in the following:

1. Decline in area of cultivated land.
2. The growing phenomenon of construction at the expense of agricultural land
3. Misuse of agricultural land and the destruction of vegetation cover.
4. Fluctuating amounts of rain and the low amount of irrigation water.
5. Poor management skills of farmers.
6. Deficient technical training of agricultural workers
7. The low quality of some of the production requirements.
8. Poor use of modern technology.

1.3.4. Impact of Human Activities

Generally, there are several established agricultural methods, six of which form the backbone of modern agriculture: application of inorganic fertilizers, irrigation, monoculture, intensive tillage, genetic manipulation of crop plants and chemical pest control. These activities have a profound impact on the environment and form a system in which each depends on the other and reinforces the necessity of using the others. The combined effects of these measures can be summarized as follows (GCEP, 2001): degradation of soil quality, contamination of soil by pesticides and inorganic fertilizers, degrading soil structure, water-logging, reduction of organic matter, salinization, soil compaction, reduction of soil fertility , soil erosion by water and wind.

Jordan is considered is third country in the world in terms of water scarcity. Wasteful use of water could be eliminated by guided agricultural practices that conserve water rather than maximize production only.

The General Corporation for the Environment Protection (GCEP) reported that the agricultural sector in Jordan has rapidly expanded during the past twenty five years. This development has been characterized by extensive utilization of chemical fertilizers, pesticides and herbicides, and the introduction of new varieties of crops. These activities aimed at increasing productivity and profitability without due consideration to their impact on soil, vegetation and the environment in general. There was no control, management or monitoring for the effects of these practices on the environment..

Intensive use of agrochemicals

Agrochemicals boost the production of agricultural crops. In their absence crops are more liable to attack and destruction by pests while their injudicious use leads to considerable problems.

The GCEP reported in 2001 that the increase in the use of nitrogen fertilizers leads to the reduction of the soil's capacity to develop and produce nitrogen from inorganic sources available naturally and metabolized by lower organisms that exchange and fix nitrogen. Farmers rely on these chemicals to obtain maximum production. Chemicals are used in agriculture at a high level without any monitoring of their impact crops or on the environment.

1.3.5. Modern Agricultural Management Techniques.

Accuracy farming is a successful agricultural management practice with the potential to mitigate difficulties and problems that face agriculture and contribute to the increase of production through utilizing of accurate information about agricultural resources. Increase of productivity depends on the wise management of input variables that may affect agriculture, management of land use, cultivation selection, plowing mechanism and irrigation techniques. Accuracy farming is based on management of land use according to soil types in different areas thus enhancing crop productivity and yield.

Several technologies are used to in accuracy farming. These include: Remote sensing, Geographical Information Systems (GIS), Global Positioning System (GPS), Change detection and monitoring of yield and vegetation, crop growth models and variable rate application. The role of remote sensing is based on monitoring specific phenomena and assessment of land covers through perfect tools as well as monitoring of the crop situation the changes it undergoes. Moreover, stored data are also employed in other hand in order to precision farming management, for instance Geographical Information Systems such as land use and land cover maps, soil types and related data, and the possibility to analysis properties of soil layers to develop applications maps for the particular field location.

It has become possible to develop precise agricultural practices or agriculture that depends on accurate allocation of areas through the combination of the Global Positioning System (GPS) and (GIS). These technologies have enabled the collection of data in actual time and the extraction of accurate information about particular sites resulting in the ability to move and analyze large amounts of data that extend across geographical space. Precise GPS data can be employed in the planning of farms, mapping of fields, preview of soil and the application of means of contrasting treatment rates and yield mapping. The system allows farmers to work during times of low visibility in the fields, as in cases of rain, dust, fog and darkness, whereby relevant data about a particular location can be retrieved repeatedly.

1.4. Water as a global agricultural problem

Water is a fundamental need and requirement all human activities on earth. Water is used in aspects of human life including domestic consumption, agriculture, and industry and energy generation. The total amount of water that resides on the surface of the earth is estimated at 1360 x 10^6 km^3. More than 97% of water exists in the oceans, and the remaining water, about 37 x 10^6 km^3 is fresh but the majority of this fresh water is of little use because it is locked in icecaps and glaciers. Groundwater on the other hand comprises about two third of the world's fresh water used for domestic, livestock watering and irrigation (Shiklomanov, I.A. and J.C. Rodda, Eds., 2003). As a result of over-pumping of surface and groundwater water resources have steadily declined. In addition, the phenomenon of global warming decreased water precipitation and increased evaporation (UNEP, 2007).

In many parts of the world the human consumption exceeds the annual limit of water, for instance in West Asia and the Indo-Gangetic Plain in South Asia. It is predicted that by 2025 almost two third of the population will be living under conditions of water stress and fifth of the global population will be living in countries or regions with absolute water scarcity (UNWater, 2007).

The water resources globally face unprecedented challenges; the most exposed areas for these challenges being the world's arid and semi-arid regions (Kundzewicz *et al.,* 2007). Population in areas of water stress are expected to increase significantly further straining the environment, coupled with the meager water resources of arid and semi-arid areas the twentieth century has witnesses an explosion of populations and expansion of unsustainable use of water and increases social expectations for domestic use of water observable in Southwestern USA for example as well as poorer countries of South America, Africa, Asia and the Middle East. (Wheater *et al.,* 2010). Jordan as one of the countries in the Middle East and North Africa region represents an excellent example of a water poor country which has seen rapidly changing demands on water resources in recent years. (Nortcliff *et al.,* 2008).

1.5. Water issues in Jordan

1.5.1. Water resources

Jordan is classified as a semi-desert area and considered one of poorest countries in the region If not globally in terms of water resource, partially as a consequence of the prevailing climatic conditions, aridity, limited amounts of rainfall, limited surface and groundwater resources and a particular demography characterized by rapid population growth and the continuing influx of refugees during the past two decades.

Processed sewage is being used at an increasing scale for irrigation, up to 75 to 80 MCM in the year 2004. Water desalination has also become an optional source; 40 million cubic meters (MCM) are currently produced, about 9 MCM for irrigation from over 10 desalination plants for the domestic supply. Renewable water sources are estimated to be around 785 to 840 MCM per annum including 280 MCM of groundwater. Surface sources provide about 505 to 560 MCM of potentially exploitable water, and non-renewable aquifers provide about 143 MCM (Duqqah *et al.,* 2007).

I. Surface Water

The public sector is the main supplier of water in Jordan. It provides water for both irrigation and domestic use. In the Jordan valley the government has established an irrigation network system to serve more than 31,174 ha of fertile valley land on the country's western border, which contributes to the total production of fruits and vegetables (USAID, 2012). **Table 1.1** surface water available to use by water resource, in addition, water coming from the wastewater treatment plants for irrigation purposes where it is mixed with fresh water to ensure dilution of pollutants (Raddad, 2005).

Table 1.1: Quantity (MCM) of surface water use by water resource 2004*

Source	Livestock	Irrigation	Industrial	Municipal	Total
1. Ground Water	**6.00**	**151.85**	**2.48**	**54.37**	**214.69**
Jordan Rift Valley	0.00	67.35	2.14	38.61	108.09
Springs	0.00	41.10	0.34	15.76	57.20
Base & Flood	6.00	43.40	0.00	0.00	49.40
2. Treated Waste Water	**0.00**	**75.40**	**0.00**	**0.00**	**75.40**
Registered	0.00	67.40	0.00	0.00	67.40
Not Registered	0.00	8.00	0.00	0.00	8.00
Total	6.00	226.25	2.48	54.37	289.09

*Source: M.O.W.I-Water Authority

The surface water by consists of fifteen basins. Surface water flow varies greatly between seasons and years. As presented in **Table 1.2,** the long term average of volume base flow is about 359 MCM/yr with about 334 MCM of flood flow. From a total annual surface water resources volume of 693 MCM/yr only 505 MCM are available for use However, considering the increase in the rate of pumping of groundwater and the decreasing flow of the Yarmouk river the estimate of 505 MCM/yr is no longer a valid figure for water resources planning purposes. The surface water available for the years 2005 and 2010 was 390 MCM and 418 MCM respectively (USAID, 2012).

Table 1.2: Long Term Average of Surface Water Flows

Basin	Base (MCM/Yr)	Flood Flow MCM/yr	Total Flow
Yarmouk	105	155	260
Jordan Valleys	19.3	2.4	21.7
North Rift Side Wadis	36.1	13.93	50.0
South Rift Side Wadis	24.8	7.7	32.5
Zarqa River	33.5	25.7	59.2
Dead Sea Side Wadis	54.0	7.2	61.2
Mujib	38.1	45.5	83.6
Hasa	27.4	9.0	36.4
Wadi Araba North	15.6	2.6	18.2
Wadi Araba south	2.4	3.2	5.6
Southern Desert	0.0	2.2	2.2
Azraq	0.6	26.8	27.4
Sirhan	0.0	10.0	10.0
Hammad	0.0	13.0	13.0
Jafer	1.9	10	11.9
Total	358.7	334.2	692.9

Source: MWI Water Budget Report, 2011

II. Groundwater

Groundwater is the main water supply source for most the population of Jordan and the regions. Jordan has been pumping from its available renewable groundwater sources at an unsustainable rate for many years, justified by the need to face water scarcity problems and to meet the demands of Jordan's growing population and economy. The Ministry of Water and Irrigation's (MWI) water budget indicates that precipitation for Jordan has fallen to 7,550 MCM in 2010 and it would be expected that recharge rates have correspondingly decreased (USAID, 2012).

The government is trying to monitor the quantity of ground water abstracted by the private sector in agriculture. Most farmers extract water for their own use directly. The mining industry also abstracts ground water for its own use. The total quantity of ground water is estimated to be 520 MCM, distributed as follow: about 54% for the agricultural sector, about 40% for the municipal sector and the remaining 6% for industrial activity, **Table 3** show the groundwater usedin MCM for a number of sources (Raddad, 2005)

Table 1.3: 2004 Ground water use in MCM*

Source	Livestock	Irrigation	Industrial	Municipal	Total
Ground Water	0.64	278.7	33.27	207.45	520.1
Renewable	0.64	210.25	29.20	192.74	432.83
Non-Renewable	0.00	68.45	4.07	14.71	87.22

*Source: M.O.W.I-Water Authority

Twelve groundwater basins have so far been identified in Jordan, listed in **Table 4**. Most of the basins have more than one aquifer. About 80% of the known of groundwater reserves are available in three main aquifers: Amman-Wadi Es Sir Aquifer System (B2-A7), Basalt Aquifer (Ba) and (D) (USAID, 2012).

Table 1.4: Groundwater Basins and Their Exploitation in 2010 (GTZ, 1977 and MWI Annual Water Budget, 2011)

Ground Water Basin	Safe Yield (MCM/Yr)	Current Abstraction	Overpumping Rate (%)	No. Of Operating wells*
Yarmouk	40.0	49.9	125.0	166.0
Side Valleys	15.0	27.7	185.0	98.0
Jordan Valleys	21.0	27.0	128.0	539.0
Amman –Zarqa	87.5	82.1	181.0	867.0
Dead Sea	57.0	90.0	158.0	327.0
Desi and Mudawarah	125*	63.2	Fossil	85.0
North Araba Valley	3.5	7.1	205.0	34.0
Red Sea\South ArabaValley	5.5	6.8	125.0	58.0
Jafer	9 (18*)	32.6	362.0	213.0
Azraq	24.0	53.19	222.0	560.0
Serhan	5.0	1.4	29.0	26.0
Hammad	8.0	11	15.0	5.0
Total	275.5	510.9	185 %	3098

That over-pumping of groundwater in some basins has exceeded the safety limit of yield by as much as three times while in others extraction of water has remained below the safety yield. Thus, the levels of depletion vary from one groundwater basin to the other. Reflecting the geographical distribution of water resources and the distribution of users. To reduce cost of infrastructure construction and immediate cost of transferring water users tend to extract water from the nearest basin resulting in heightened pressures on water sources (Raddad, 2005). **Table 1.5** illustrates ground water availability and extraction from each ground water basin.

Table 1.5: Groundwater resources and Use 2004

Ground Water Basin	Safe Yield (M.C.M)	Water usage (M.C.M)					Balance	% Safe Yield
		Municipal	Industry	Agriculture	Distance Regions	Total Use		
Yarmouk	40.0	0.5	0.1	35.0	0.0	43.3	-3.3	108.0
Side Valleys	15.0	0.0	0.0	3.3	0.0	25.9	-10.9	172.0
Jordan Valleys	21.0	0.0	0.2	19.6	0.0	27.9	-6.9	133.0
Amman -Zarqa	87.5	7.7	5.7	54.7	0.0	138.7	-51.2	158.0
Dead Sea	57.0	2.1	12.5	30.1	0.2	89.3	-32.3	157.0
Desi and Mudawarah	125*	0.0	4.1	68.4	0.0	82.1	42.9	66.0
North Araba Valley	3.5	0.0	3.3	2.8	0.0	6.7	-3.2	193.0
Red Sea\South ArabaValley	5.5	0.0	0.4	15.9	0.0	17.4	-11.9	316.0
Jafer	9 (18*)	0.4	6.8	10.7	0.0	14.8	-15.8	276.0
Azraq	24.0	0.1	0.2	34.4	0.1	59.3	-35.3	247.0
Serhan	5.0	0.0	0.0	3.6	0.2	3.8	1.2	76.0
Hammad	8.0	0.0	0.0	0.0	0.2	0.9	7.1	11.0
Total	275.5	6.9	33.3	278.7	0.6	520.1	-170.8**	

*Source: M.O.W.I-Water Authority

* Nonrenewable Ground Water

** Sum of Renewable Ground Water

1.5.2. Water Quality

Studies of water scarcity in Jordan are not limited to the question of restricted quantities but also address water quality issues Jordan's natural, ecosystems and aquifers are subject to increasing levels of pollution, evident in the Zarqa River, the Amman-Zarqa aquifer as well as the downstream Jordan valley and the lower Jordan River. Flow along the Yarmouk basin declined by 10 percent from the level of mid 1990s with associated deterioration in water quality. Sewage effluents and saline springs some brackish base flow into the lower Jordan River. In short, Jordan's surface water supplies are declining in terms of both quantity and quality (USAID, 2012). There is a crucial need to focus on increasing supply and expanding treatment of sewage effluents so that wastewater could be recycled and enter as an effective source in the water balance. Treatment reduced the microbiological content of sewage effluents making wastewater more usable but it does not reduce salts. Levels of total dissolved solids (TDS, salt) increased in Amman-Zarqa basin like As-Samra plant to about 1,165 ppm TDS and also by effluent. This increase in salinity is attributable to industrial installations.

In addition to strategies that ensure an adequate quality of water for projected uses and meaningful guidelines regarding the quality of water in agriculture are required. It is also necessary to enforce existing national health standards for municipal water quality (JISM, 2001). The guarantee of water quality is essential element of the overall strategy for water resources. There is a pressing need to prevent the pollution of water at the surface through inappropriate human activities such as the excessive application of fertilizers and pesticides and to guard the quality of groundwater against decline (Nortcliff *et al.*, 2008).

1.6. Objectives of the Research Work

In an attempt to test the potential of using precision geoinformatics techniques to improve the management of the Jordanian agricultural sector and the monitoring of water sector the following objectives were pursued in this study:

1. Investigation of land degradation problem by geoinformatics/ if there is a land degradation problem:
    - Validation of the impact of human activities on land degradation
    - Mapping of land degradation

2. Identification and Mapping Groundwater Vulnerability
3. Prediction and evaluation of the quality of groundwater
4. Investigation of the relationship between vulnerability of groundwater and land degradation through hot-spot areas.

I. Research questions

To fulfill these objectives the following research questions were framed:

For First objective,

1. Could land degradation be verified using geoinformatics?
2. How do agricultural human activities affect in the degradation of the vegetation and the deterioration of the quality of the soil?

For second objective,

1. How can the vulnerability of groundwater be validated and modelled.

For Third objective,

1. Could SAR be considered an influential factor on deterioration of groundwater quality?

For Fourth objective,

1. Is there a relationship between vulnerability of groundwater and land degradation and what is the appropriate method to map this relationship?

II. Brief summary

To achieve these objectives a number of actions were carried out based on the available data: as a first step validation and modeling of groundwater vulnerability using the DRASTIC model followed by a modified DRASTIC model with land use mapping using GIS techniques; as a second step values of SAR were measured in the wells to validate the impact of SAR on the deterioration of groundwater quality.; as a third step field data and remote sensing data was collected to verify land degradation using geoinformatics techniques and the development of a respective model where the degree if land degradation appears is mapped ; as a fourth step the intersection between vulnerability of groundwater results and land degradation results was explored to identify hot-spots a using geoinformatics techniques.

1.7. Structure of the thesis

This thesis contains an introduction and six chapters as follows:
The introduction has discussed Agriculture and water issues in Jordan which carry the content of a number of problems including resources, demands and quality, as well as outlined the scope of this research and its aims.

Chapter 2, Literature review:

This chapter reviews the concept of Land degradation, including applications of geoinformatics in Land degradation and methods of assessment, and the different parameters affecting soil fertility and agriculture production, gives some fundamental information on soil and water models and water quality methods.

Chapter 3, study area and Material:

The third chapter describes the study area, including the selection criteria, and the development of an inventory of available secondary data. It will also describe the study area characteristics in NE Jordan with regard to climate, rainfall, geology, hydrology, geomorphology, soils, and groundwater, the chapter also includes an explanation of the criteria of the sources and dataset used.

Chapter 4, Methodology:

The methodology chapter contains section about remote sensing processing, validation of the data that was collected and image classification with also explain the methodology for change detection that appeared in land use in the study area twenty years ago, and one of the other sections includes and describes the methodology for fieldwork and labs analyses of groundwater hydrochemistry for 2012 and 2013, and the soil chemistry data (geochemistry), and will study which involves descriptive statistics as well as section about processes geoinformatics data on the environment of the study area that includes methods development of groundwater vulnerability and land degradation with method of integration.

Chapter 5, Results: In this chapter, the results obtained from the evaluation of the data are presented: This consists of the extraction of thematic information from the satellite imagery and the detection of change from the time series data. Further analysis of the soil and water samples and the subsequent statistical analysis were done to determine the parameters that used in modified of groundwater vulnerability index (DRASTIC modified) and develop an ESAs model (Environmental Sensitive Areas model) to a new degradation Model (DM) model with a new formula and validation

Chapter Six, Discussion, Conclusions and Recommendations this chapter will consist of a discussion of the material, methods and results for this research, a summary of the principal research findings, the limitations of the study and recommendations.

Chapter 2

Review of Literature

2.1. Introduction

As discussed in the introduction, water is a fundamental requirement for all human activities, and these activities have a major impact on agriculture, which is a basic pillar of economic and social development in all countries in general on great impact on vegetation and in particular on water quality. Human activities and agricultural practices can also have adverse environmental effects, with soil degradation being the most dangerous, most likely, and most direct implication, next to the pollution of soil, water, and air, the result being the deterioration of the environment as a consequence of inappropriate agricultural practices. This review covers the main factors leading to soil degradation as well as the monitoring techniques and modeling methods of water quality and soil.

2.2. Land Degradation

Land degradation is a process that leads to a decline in soil productivity at the level of current and/or potential capacities; it is also defined as a decline in the ability of land to perform ecosystem goods, services, and functions that support and develop society (MEA, 2005; FAO, 1979). Both ecosystem degradation and economic loss could be considered as measures for the ability of the land to meet human needs (Kassas, 1995). Land degradation has also been defined as the decline in the physical and chemical characteristics of the soil as a result of environmental change (Imeson & Emmer, 1992). Land degradation is the result of natural operations of human activity where the land is no longer able to support the economic field correctly and/or its original natural ecosystem functions. Degradation often leaves clear implications on land through land use and land-cover change (Abbas and Fasona, 2012). Considering all these concepts, it could be concluded that land degradation is one of the most dangerous environmental problems in the world: it involves two interrelated and complex systems: the natural ecosystem and the human social system, whereby biophysical processes do not represent all causes of land degradation; these include socio-economic factors (i.e. human health, institutional support, poverty, marketing, income), as well as factors conditioning political stability that influence food production (Fadhil, 2009; WMO, 2006; UNCCD, 2004).

The degradation of land in arid, semi-arid, and dry sub-humid areas results from various factors, including climatic variations and human actions that accelerate desertification (UNCED, 1992). It should be highlighted that overgrazing and agricultural practices are considered to be the main principle drivers of land degradation in arid and semi-arid regions. It involves two main processes: soil erosion and loss of soil by wind and/or water. However, salinization, seawater intrusion, and water logging also play a role, especially in irrigated coastal plains, due to the depletion of groundwater and inappropriate practices. The deterioration of the physical, chemical, biological, and economic properties of soil are further reasons for land degradation next to deforestation and of the long-term loss of natural vegetation (ESCWA, 2007). Soil degradation is considered to be a crucial component of land degradation; land degradation includes water and wind erosion, chemical, physical, and biological degradation as well as salinization (Mhangara, 2011).

Land-degradation is significantly affected by land use/land cover changes such as soil erosion; they are an important input variable into research on global change (Li *et al.*, 2009; Verburg *et al.*, 2009). In addition, land use and land-cover changes respectively are considered to be the main reasons for soil degradation that directly affect ecosystem services supporting human needs worldwide. Land use and land-cover change have a significant effect on major ecological functions. These processes are also considered to be among the most critical issues for global environmental change (Dong *et al.*, 2009; De Chazal and Rounsevell, 2009). Human practices and activities (anthropogenic impacts) significantly modify patterns of land-cover change and increase segmentation of the landscape (Muriuki *et al.*, 2011). The world has identified the environmental challenges of agricultural land use, soil degradation, and erosion as probably the most significant environmental problems, whereby unsustainable farming practices and land use, including mismanaged intensification and land abandonment, have an adverse impact on natural resources (Louwagie *et al.*, 2009).

Land degradation could pose a major threat to food production systems and rural livelihoods, particularly if the expansion of agricultural production in marginal lands continues at the current pace. With the rapidly increasing water needs of other agricultural sectors and the continued inefficiency of on-farm water use, the quantity and quality of water will be threatened, which would imply having to give up agricultural land. Corresponding to the several types of land degradation, there are multiple paths to improve land. Improve land investments and better land management practices through appropriate policies can promote the conservation of the resource base through many programs by well-targeted public investments. . Taking in consideration what has been implemented in the context of the comprehensive policy and supportive to elimination of discrimination against marginal areas, promotion of rural income growth and diversification, correction of market distortions, promote rural development and institutional innovations (ESCWA, 2007).

There is a global trend towards monitoring land degradation in a consistent way, but there is little detail on the immediate processes involved, whether it is datasets or statistical analysis. New models, model integration, enhanced statistical analysis, and modern sensor imagery at medium spatial resolution should substantially improve the assessment of global land degradation (Jong *et al.*, 2011).

2.3. Soil Fertility and Agricultural Production

Properties and variability of soil parameters have been analyzed and documented in many studies by using a variety of scientific methods (Havlin *et al.*, 1999; Jones, 1998; Logsdon *et al.*, 1998; Mills and Jones, 1996 and Bergmann, 1992). The parameters of soil documented in the literature are generally based on major aims stipulated in the respective research studies. In order to achieve the goals of this research, we analyzed parameters of the soil that have an impact on the productivity of agriculture, composed organic matter (OM), Sodium-Absorption Ratio (SAR), PH, and electrical conductivity (EC).

I. Organic Matter (OM)

Soil organic matter is carbon-rich and typically contains about 50% carbon. The remainder of soil organic matter consists of about 40% oxygen, 5% hydrogen, 4% nitrogen, and 1% sulfur. Its material contains living biomass (like plants, animal tissues, and microorganisms), dead tissues (i.e. partially decomposed materials) and non-living materials, namely stable parts formed from decomposed materials also known as 'humus' (Toor and Shober, 2008).

Soil OM supports the functions of the soil's environmental quality because it links particles of the soil together into stable aggregates, thus improving porosity, infiltration, root penetration, and reducing runoff and erosion (Christensen and Johnston, 1997). Furthermore, it enhances soil fertility and plant productivity by improving the ability of the soil to store and supply nutrients, water, and air (Zanen and Koopmans, 2005). Soil OM also provides food and habitat for soil organisms. Carbon isolates from the atmosphere reduces runoff and mineral crust formation and consequently reduces the negative impact on water quality and the environment (Herrick *et al.*, 2001). Soil organic matter binds selected harmful pollutants like residual pesticides and trace elements so that these constituents cannot escape from the soil and pollute our water bodies (Toor and Shober, 2008).

II. Soil pH

Soil pH is a measure of the soil solution's acidity and alkalinity. A pH of 7 is neutral, a pH below 7 is acid, and a pH above 7 is alkaline (Anne *et al.*, 2009). The soil pH influences crop yields, plant-nutrient availability, and microorganism activity- factors that have an impact on key soil processes. On the other hand, there are inherent factors affecting soil pH, such as mineral content, climate, and soil texture. Natural soil pH reflects the combined effects of soil-forming factors (parent material, time, topography, climate, and organisms). Soil pH can be managed by measures such as applying the proper amount of nitrogen fertilizer, liming, and cropping practices and cropping practices that improve percentage of organic matter and overall soil health Soil pH is modified through land use and management. Vegetation types impact on the soil pH. Several factors lead to changes in the land after a few years of use, such as the loss of organic matter, removal of soil minerals when crops are harvested, erosion of the surface layers, and effects of nitrogen and sulfur fertilizers.

As a result of these factors, soil pH in forestlands tends to be more acidic than in areas of grassland, a factor to consider in the conversions of forestland or grassland to cropland (NRCS, 2012). Measurements of pH are thus an important component of soil-analysis procedures. **Figure 2.1** illustrates the impact of pH on microbial activity and nutrient availability in mineral soils.

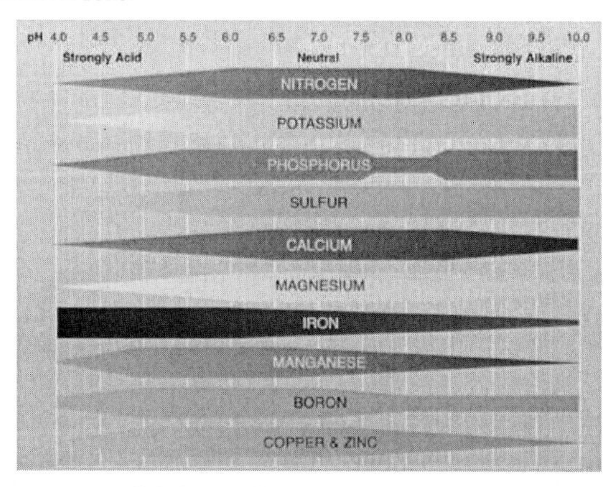

Figure 2.1: Nutrient availability and microbial activity as affected by soil pH; the wider the band, the greater the availability or activity (adapted from Truog, USDA Yearbook of Agriculture, 1943-1047).

III. Sodium-Absorption Ratio (SAR)

SAR is an index that expresses the proportion of exchangeable sodium (Na^{+}) to exchangeable calcium (Ca^{2+}) and magnesium (Mg^{2+}) ions. The formula to calculate SAR is given below in **Equation 2.1**, with concentrations expressed in milliequivalents per liter (meq/L) and analyzed from a saturated paste of soil extract (Sonon *et al.*, 2012). When SAR values more than maximum, the soil is called a *sodic soil*. Increasing sodium in sodic soils leads to the strong agglutination of soil particles preventing the formation of soil aggregates. This results in a very tight soil structure with poor water infiltration, poor aeration, and surface crusting, which makes tillage difficult and restricts seedling emergence and root growth (Munshower, 1994; Seelig, 2000; Horneck *et al.*, 2007). This can reduce plant growth within a few years. **Figures 2.2** show the impact on plant growth over time. With the help of **Table 2.1**, it is possible to determine the status of soil based on its values of SAR and thus of soil quality to sustainability of plant growth.

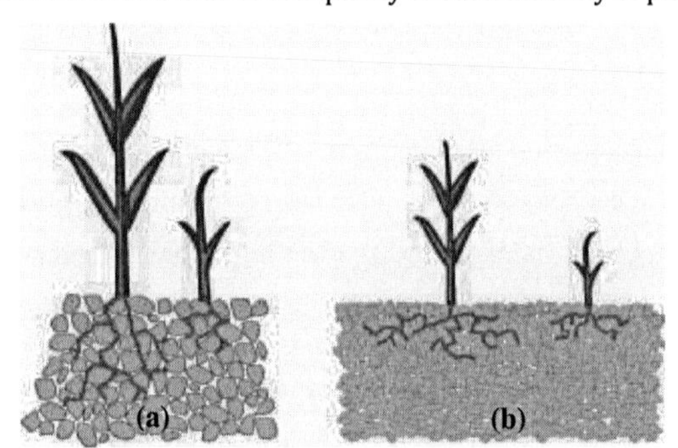

Figures 2.2: (a) Soil with good structure (non-sodic soil). (b) Soil with poor and dense structure (sodic soil) (*Sonon et al., 2012).*

$$SAR = [Na+] / \sqrt{[Mg+]+[Ca+]/2} \qquad 2.1$$

Table 2.1: Soil-Quality Guideline for SAR and EC Values in the Unrestricted Land Use (Alberta, 2001).

Parameters	Rating Categories			
	Good	Fair	Poor	Unsuitable
EC dS/m (Salinity)	< 2	2 - 4	4 - 8	> 8
SAR (Sodicity)	< 4	4 - 8	8 - 12	> 12

IV. Effects of Electrical Conductivity (EC) on Yield

EC measures the ability of the soil solution to conduct electricity and is expressed in decisiemens per meter (dS/m, which is equivalent to mmhos/cm) (Sonon *et al.*, 2012). Extraction of the spatial measurements of the soil's electrical conductivity (ECe) is useful for characterizing spatial variability because ECe is influenced by several soil properties (pH, moisture content, texture, topography, depth of ground water, etc.) (Corwin *et al.*, 2003; Kitchen *et al.*, 2003). **Table 2.1** helps clarify the impact on soil and is also useful for devising sampling schemes in order to identify soil properties that influence yield in instances where ECe correlates with crop yield and maps of ECe (Heiniger *et al.*, 2003).

2.4. Geoinformatics in Land Degradation

Geoinformatics of land degradation may be evaluated qualitatively or quantitatively. The first method relies on expert opinion in the evaluation of the effects and symptoms of degradation, such as decline in biomass, vegetation health, and land quality. Qualitative approaches focus on the sensitivity of land-mapping units to external changes that impose limitations on the farmers' freedom of action and require good knowledge of the capacity and classification of the earth in terms of its capability (Klingebiel & Montgomery, 1961; Hockensmith & Steele, 1949).

Quantitative methods employ proxy measures like spectral reflectance. Most frequently, remote sensing that depends on establishing relationships between the proxy and the 'real thing' is used, and according vegetation dynamics that provide repeatable and transferable features between regions and scales, it is easier to use remote sensing where land degradation processes impact on the vegetation cover. Remote sensing has been adopted widely as well as on a global scale as an indicator of land degradation In an assessment of land degradation, the approach depends on classes applied to a common base of landform units, depending on the degree (light – severe) and the frequency (percentage occurrence within the mapping unit) of degradation reason, nutrient depletion, soil erosion, salinity and/or chemical contamination (Jong *et al.*, 2011). Algorithms are used to calculate classes combining loss of vegetation as well as a decision tree are employed to derive a land-degradation map (Vargas *et al.*, 2007).

Several criteria have been approved for monitoring the ecological status such as statistical rigor, reproducibility, scalability, transferability, and cost-effectiveness (Boer, 1999). Geo-information technology provides and meets many of these criteria.

Integration of geo-information technology (Remote Sensing („RS"), Geographic Information Systems („GIS"), and Global Positioning Systems („GPS")) makes it a basic and essential technical core of the system of geo-space information science that plays an important role for evaluating, monitoring and investigating the environment and its components, it is used to determine is used to determine land condition trends (whether conditions are getting worse, better, or stay the same) and is particularly beneficial for monitoring desertification and land changes over time (Fadhil, 2009).

2.5. Remote Sensing and Image Interpretation

The overall goal of image classification of RS data is the extraction of information contained in the images, depending on the several spectral or terrestrial characteristics of the earth's surface and its electromagnetic spectrum (Lillesand and Kiefer, 2000; Murai, 1996). Through remotely sensed imagery, it is possible to classify land use and cover by using two general image analysis approaches: i) classifications based on pixels, and ii) classifications based on objects (Duro *et al.*, 2012).

2.5.1. Pixel-Based Image Analysis

This approach relies on the categorization of individual pixels. Only the spectral patterns are used, without taking into account the neighborhood or spatial information of the pixel (Lee and Warner, 2004). The classification manner is (pixel by pixel) and any one pixel can only belong to one class. Pixel-based classifications are traditionally either supervised or unsupervised.

Supervised and Unsupervised Classification

The objective of these processes is to categorize all pixels into a digital image according to land-cover classes or themes. These categorized data may then be used to produce thematic maps of the land cover presented in an image and/or as summary statistics of the areas covered by each land-cover type. Visual analysis of the image data can be replaced with quantitative techniques for automating the identification of features in a scene. This normally involves the analysis of multispectral image data and the application of statistically based decision rules for determining the land-cover identity of each pixel in an image (Lillesand and Kiefer, 2000). Based on data, file values in individual classes or categories of data for the pixels of objects are classified by using one out of three approaches: the supervised, the unsupervised, or the hybrid approach based on the spectral classification of images.

2.5.1.1 Supervised Classification

Supervised classification depends on the feature that each spectral class can be described with a probability distribution in multispectral space. Each distribution describes the chance of finding a pixel belonging to that class at any given location in multispectral space (Richards and Jia, 2006). The analyst relies on his/her own pattern-recognition skills and priori knowledge of the data in order to help the system determine the statistical criteria (signatures) for data classification. These signatures drive the assignment of the remaining unlabeled pixels in the image (Ngamabou, 2006). A priori knowledge refers to the acquisition of information about the study area from field work, analysis of aerial photographs, personal experience, etc. These data are considered to be the most accurate about the area and are generally referred to as (ground-truth data). The classification accuracy is not overly sensitive to violations of the assumptions that the classes are normal. But as generally, a vector layer consisting of various polygons overlaying the different classes in mind (land-use types or classification scheme) is digitized over the raster scene. Supervised classification procedures are essential analytical tools used for the extraction of quantitative information from remotely sensed image data.

A) Training Stage

The main objective in this operation is to collect a set of statistics that reflect spectral response patterns in the image for each land-cover type classified. Training samples are set points describing each category of land cover, the analyst selects training samples in order to train the classifier that represents typical spectral information of land-cover classes. All spectral classes constituting each information class must be adequately represented in the training-set statistics used to classify an image (Lillesand and Kiefer, 2000).

B) Classification Stage

Classification is carried out after determining a set of training samples and a specific classification algorithm where all pixels in the image are numerically compared to the training samples that reflect the land-cover classes in accordance with specific algorithms. The classifiers that are generally used are: minimum distance to mean classifier, parallelepiped classifier, and maximum likelihood classifier.

1. *Minimum Distance*

The spectral value in each band is determined for each information class. These spectral values include median vectors for each class. A pixel of unknown identity may be classified through calculation of the distance between the value of the unknown pixel and each of the class means. If the pixel lies further away from any class mean than the analyst-defined distance (distance threshold), it is classified as "unknown". In the minimum distance technique, class models are symmetric in the spectral domain (Richards and Jia, 2006; Lillesand and Kiefer, 2000).

2. *Parallelepiped Classifier*

The parallelepiped classifier is similar to the minimum-distance algorithm but is sensitive to variance by integration of the range of spectral values in each category training set. An unknown pixel is classified according to the category range or decision region (Lillesand and Kiefer, 2000).

3. *Maximum Likelihood*

Maximum likelihood classification is the most common supervised classification method used with remote-sensing image data. This is developed subsequently in a statistically acceptable manner; where the maximum likelihood approach is based on comparisons of the upper and lower limits of data values for all candidate-unit pixels, and in relation to the minimum and maximum data-file values of each band in the signature; the mean of each band, plus and minus a number of standard deviations or limits specified by the analyst and based on his/her knowledge of the data and signatures. The principle drawback of maximum likelihood classification is the large number of computations required to classify each pixel (Lillesand and Kiefer, 2000).

2.5.1.2. Unsupervised Classification

Unsupervised classification operates through appointing the pixels of the image to the spectral classes without the user having foreknowledge of the existence or names of those classes. It is performed most often by using clustering methods. These procedures could be used to identify the number and location of the spectral classes in the data and to determine the spectral class of each pixel.

The analyst can identify these classes by linking a sample of pixels in each class with available reference data afterwards. This could include maps and information from ground surveys. The analyst is usually totally unaware of the spectral classes or sub-classes as they are sometimes called (Richards and Jia, 2006). In this case, the analyst is required to interpret spectral clusters created by the unsupervised training and to include in algorithms in hard information classes (Ngamabou, 2006).

Clustering procedures are usually mathematically expensive but considered to be central to the analysis of remote sensing imagery when the information classes for a particular exercise are known. Unsupervised classification is therefore useful to determine the spectral class composition of the data prior to detailed analysis by supervised classification. The general objective is to develop approaches of image classification over the years in response to the need for more accurate results (Baatz, 2004; Yu, 2006).

2.5.2. Object-Based Image Analysis

This methods relies on the categorization of the whole image from group pixels into objects through an image-segmentation process based on a chosen similarity (e.g. texture, color, intensity) - and then the use of the spectral, spatial, and contextual information inherent in these objects (Malambo, 2009; Navulur, 2007; Baatz, 2004). It is difficult to exploit the specialized knowledge and contextual information adequately with pixel-based classifications (Flanders *et al.*, 2003). Object-based image-processing techniques have the ability to extract real-world objects proper in shape and accurate in classification (Baatz, 2004). By segmenting the image into multiple pixels, object-based analysis provides an indication of various sizes depending on spectral and spatial characteristics of pixels groups .

2.5.3. Accuracy Assessment

Accuracy refers to the level of compatibility between the signs (as assigned by the analyst) and locations of classes on the ground that they were collected by the user as test data In order to obtain valid conclusions about a map's accuracy from samples of it, they must be selected without bias because the places assigned to meet these important criteria affect the validity of further analyses that could be carried out by using the data, which negatively affects estimates of true accuracy.

Overall Accuracy

Overall accuracy is a very coarse measurement. It gives no information about which classes were classified with good accuracy. The procedure "overall accuracy" reflects the proportion of all reference pixels that are classified correctly. It is computed by dividing the total number of correctly classified pixels (the sum of the elements along the main diagonal) by the total number of reference pixels (Congalton, 1991). The overall accuracy can be calculated according to the following equation:

[overall accuracy = (total number correct) / (total reference or total classified)* 100]

Producer's Accuracy

This measure is called “producer's accuracy” because the producer of the classification is interested in how well a certain area can be classified. In this procedure, the total number of correct pixels in a category is divided by the total number of pixels of the category that was derived from the reference data. This accuracy measure indicates the probability of a reference pixel being correctly classified and really is a measure of omission errors (Congalton, 1991).

The producer's accuracy can be calculated by using the following equation: (producers accuracy = number correct / reference total)

User's Accuracy:

The measure "user's accuracy“ is a map-based accuracy calculated by looking at reference data of classes and then determined by the percentage of correct predictions for these samples. User's accuracy is estimated by dividing the number of pixels of the classification result for class A by the number of pixels that agree with the reference data in class A. The measure is an estimate of the probability that a pixel classified as class A actually does belong to class A (Story and Congalton, 1986). The user's accuracy can be calculated by using the following equation:

(user's accuracy = number correct / classified total)

2.5.4. Digital Change Detection

Digital change detection is a technique used in remote sensing to monitor and manage natural resources. It provides quantitative analysis of spatial distribution in the area of interest. The method is useful in several applications: deforestation assessment, vegetation phenology, urban expansion and planning, damage assessment, crisis management and response, crop-stress detection, snow melting, etc. (Tardie and Congalton, 2002).

In addition to determining the changes and differences in a particular object or phenomenon of study through monitoring it at different times (Singh, 1989), the process of obtaining remote sensed data from Earth-orbiting satellites is repeated at varying intervals (temporal resolution) in order to detect changes. Several approaches were developed to detect changes in land cover, for instance traditional post-classification cross-tabulation, cross-correlation analysis, neural networks and knowledge-based expert systems – with most of the procedures relying on pixel-based methods (Muttitanon and Tripathi, 2005). These methods are commonly used with moderate resolution images and not the high spatial resolution of "new" remote sensing images (Civco *et al.*, 2002). Post-classification change detection is widely used and the direction of change is also provided for in other methods (Peterson *et al.*, 2004).

Change detection in two images between two different dates is considered to be changes in radiance. Different factors can induce changes in radiance between two dates, such as changes in sensor calibration, solar angle, atmospheric conditions, seasons, or Earth surface. Actual change in the Earth surface is detected when changes in radiance are identified by using values produced from imagery for change detection of the Earth surface.

Moreover, the change in radiance as a result of the changes that occur on the surface of the Earth must be large in comparison to the change in radiance due to other factors

(Théau, 2012). In order to detect changes of the Earth surface by using imagery, various image dates must be selected and compared to each other carefully. The selection of data is a critical step to study change detection. The acquisition period (i.e. season and month) of different dates for the images is an important parameter to be considered in the method of image selection because it is directly related to phenology, climatic conditions, and solar angle (Théau, 2012). In order to minimize the effects of these factors, great care must be taken in the selection of multi-date images.

In order to compare the images that are classified and analyzed between two periods in order to determine the change-detecting matrix, the images must be drawn from the two periods of time and have to be classified separately. Different particles on the soil surface must be suitably labeled. Through these steps, it is then possible to construct the change map (Churchill, 2003).

Nevertheless, caring for the data selection exclusively is not enough to reduce radiometric heterogeneity between multi-date images; there are also other criteria that must be considered for correction, such as atmospheric conditions and solar angle differences as well as sensor calibration and geometric distortions (Théau, 2012). Comparison of the multi-date images is normally performed on the basis of pixels. Accordingly, these images tend to have very accurate registrations in order to compare pixels from the same locations. An error in the registry of multiple image dates can cause significant errors in the interpretation of changes (Wang and Ellis, 2005). Methods available and used in change detection studies have become commonplace, such as image differencing, image rationing, post-classification, direct multi-date classification, linear transformation, change-vector analysis, image regression, multi-temporal spectral mixture analysis, and combined approaches (Coppin *et al.*, 2004; Lu et al., 2004; Mas, 1999; Lunetta and Elvidge, 1998; Mouat et al., 1993; Singh, 1989).

2.6. Issues of Image Processing

Remotely sensed data (when obtained from imaging sensors mounted on satellite platforms) generally contain deficiencies and flaws. They contain errors in geometry and in the measured brightness values of the pixels. The latter are called "radiometric errors" and can result from the instrumentation used to record the data, from the effect of the atmosphere, and from the wavelength dependence of solar radiation. At this stage, the data are referred to as "pre-processed", pending the correction of these flaws and deficiencies in order to obtain more accurate results and to eliminate exogenous differences. The preprocessing activities are identified by the user based on the objectives pursued.

The most difficult issue in the all-monitoring processes are operations subject to monitoring for a period of time. The use of multi-date images creates a degree of uncertainty in mapping the changes, knowing that measurements in the dataset change over time.

I. Geometric Correction

Geometric enhancements are features of remote sensing generally related to smoothing, edge detection and enhancement, as well as line detection. Enhancement of edges and lines leads to image sharpening (Richards and Jia, 2006). Information derived from remote sensing images is most often incorporated with map data in a GIS or is presented in the form of maps. Images of remote sensing are not maps and thus need to be converted in accordance with the scale and projection of maps in a process referred to as "geometric correction" (Mather, 1999).

Image geometry errors can occur in many ways. The relative motions of the platform, its scanners, and the Earth, for example, can lead to errors of a skewing nature in an image product. The curvature of the Earth and uncontrolled variations in the position and attitude of the remote sensing platform can all lead to geometric errors of varying degrees of severity (Richards and Jia, 2006).

Various types of geometric distortion can be processed by using two corrective techniques. The first involves modeling the nature and magnitude of the sources of distortion and using these models to establish correction formulae. This technique is effective when the types of distortion are well characterized, such as those caused by Earth rotation. The second type is based on establishing mathematical relationships between the addresses of pixels in an image and the corresponding coordinates of these points on the ground (via a map) (Richards and Jia, 2006).

II. Radiometric Correction

In contrast to geometric correction, in which all sources of error are often rectified together, radiometric correction procedures must be specific to the nature of the distortion. Two broad types of geometric distortion result from measured brightness values of an image's pixels. In the first type, the relative distribution of brightness over an image in a given band can be different to the one in the ground scene. In the second type, the relative brightness of a single pixel can be distorted from band to band and compared with the spectral reflectance character of the corresponding region on the ground. Thus, both types can result from the presence of the atmosphere as a transmission medium through which radiation must travel from its source to the sensors – and can also be a result of instrumentation effects (Richards and Jia, 2006).

Generally, atmospheric attenuation and topographic distortion (sensor variation) are considered to be the main factors that seriously affect the radiometric quality of digital satellite data. Other factors like illumination angle and soil moisture also affect radiance values as well as (consequently) classification results and change information. These factors need to be either corrected or taken into consideration.

2.7. Methods of Modeling Surface and Subsurface

The application of detection-change techniques remains restricted to controlling and monitoring the effects of the Earth's surface. Geoinformatics methods (GPS, GIS and remote sensing) could also be used to monitor the particular case (such as groundwater vulnerability assessment, following the pathway of pollutions in the soil) under the Earth's surface or to follow a serious impact on the environment (like land degradation, soil erosion, desertification) by using a variety of modeling methods.

2.7.1 Groundwater-Vulnerability Assessment Methods

The concept of groundwater problems reflects global concern about the quality of groundwater. Natural deterioration in it is rare. Usually, deterioration comes about as a result of human activities such as the intrusion into groundwater of salt water or water unsuitable for human consumption as well as direct pollution through leaching of wastes and chemicals from the land surface or underground sources vertically down into an aquifer (SNIFFER, 2004). The concept of groundwater vulnerability rests on the assumption that the physical environment may provide some degree of protection to groundwater against natural and human impacts (Abdullahi, 2009).

A description of the degree of groundwater contamination as an indicator of the hydrogeological conditions that surround a specific area can be presented through the construction of a typical map. This indicator depends on various attributes that reflect hydrogeological conditions such as vadose zone characteristics, aquifer depth, the amount of recharge, and other attributes.

Maps of groundwater vulnerability provide information about the degree of pollution and reflect the impact path of pollutants resulting from these activities. Water tables are used to benefit from these maps in the assessment of the risks and consequences that occur due to contaminants (SNIFFER, 2004). Monitoring the pathway of these pollutants in the soil brings the attention of land-use planners to existing problems and helps in the development to modify the potential occurrence of harmful conditions like groundwater contamination before causing serious harm (Murray and Rogers, 1999). Vulnerability-mapping techniques are used to classify geographical areas with regard to their susceptibility to groundwater contamination (Knox *et al.*, 1993). Differentiation between one area and another is based on several physical and environmental factors (Barry *et al.*, 1990).

Several techniques have been developed to assess groundwater vulnerability. The most common ones are the DRASTIC system (Aller *et al.*, 1987), the GOD system (Foster, 1987), the AVI rating system (van Stempvoort *et al.*, 1993), the SINTACS method (Civita, 1994), the ISIS method (Civita and De Regibus, 1995), the Irish perspective (Daly *et al.*, 2002), the German method (von Hoyer and Sofner, 1998), and EPIK (Doerfliger *et al.*, 1999). These techniques are based on hydrogeology parameters by combining maps of parameters considered to be effective factors in contaminant transport, such as soil, geology, land use, depth to groundwater table, hydraulic aquifer conductivity, and stream network – whereby each parameter is assigned a numerical score different in each technique and based on its perceived importance. The assignment of an equal score to each parameter – regardless of its importance – is considered to be the simplest method. Many other quantitative methods tend to have different numerical scores and weights for each parameter in order to indicate the impact degree of each parameter on groundwater vulnerability in a region (Evans and Maidment, 1995).

The most renowned of the above-listed methods is DRASTIC (Aller *et al.*, 1987). It could be used in intergranular porosity and limestone aquifers. It is considered to be one of the most frequently applied methods on the international level. The DRASTIC methodology was developed by the US Environmental Protection Agency (Aller *et al.*, 1987). The model combines parameters through spatial datasets on **D**epth to groundwater, **R**echarge by rainfall, **A**quifer type, **S**oil properties, **T**opography, **I**mpact of the vadose zone, and the hydraulic **C**onductivity of the aquifer (Engel *et al.*, 1996). Determination of the agricultural DRASTIC index involves multiplying each factor's weight by its point rating and summing up the total (Knox *et al.*, 1993). The overall 'DRASTIC index' is established by applying the following formula **2.2**:

$$\mathbf{DI = DrDw + RrRw + ArAw + SrSw + TrTw + IrIw + CrCw} \qquad \mathbf{2.2}$$

Where:
r: rating for the area being evaluated.
w: importance weight for the parameter.
D: Depth to groundwater.
R: Recharge rate (net).
A: Aquifer media.
S: Soil Media.
T: Topography.
I: Impact of the vadose zone.
C: Conductivity (hydraulic) of the aquifer.

Considering the advantages and limitations of the methods outlined in this review, the most suitable combination of techniques to investigate groundwater quality in this study is arguably GIS together with remote sensing. Necessary data for the mapping of land use were collected through remote sensing while spatial and modeling approaches to investigate groundwater quality were performed with GIS. Sodium-absorption ratios from wells irrigating agricultural land could be calculated by using a GIS based model. Groundwater vulnerability was evaluated through the integration of GIS, remote sensing and DRASTIC. DRASTIC was chosen for the following reasons:

1. The availability of data: The available data for this project was not sufficient to use any mathematical or statistical model recommended in the literature. Such methods require large amounts of data collected over a long period of time.

2. The use of GIS: DRASTIC is known through the published literature cited in this chapter to be easily implemented entirely in a GIS environment due to the existence of the overlay procedure and the arithmetic operation functions in various GIS software, which facilitates the implementation of such a model.

2.7.2 Soil Degradation Assessment Methods

Environmental degradation, especially in arid and semi-arid lands – including deforestation, land degradation, soil-quality degradation, erosion soil, and desertification, – is a cause for global concern. Degradation has increased in recent years as a result of human activities such as using salt water for irrigation, intensive use of biochemical materials, weakness of agricultural management, as well as negative farming practices (Rodriguez *et al.*, 2004). Monitoring environmental degradation types through modeling methods might provide some degree of protection against natural and human impact. Generally, land degradation is triggered by detrimental human actions which impacting with extreme and persistent natural forces that stress ecosystems (WMO, 2005).

The interaction of these forces determines the intensity of degradation. Farming practices and human activities can be viewed as bio-geophysically and socioeconomic-politically (Gad, 2008). Land use and land management, climate, biodiversity, terrain and soil type are considered to be bio-geophysical factors; as well as land tenure, institutional support, income, human health, incentives, political stability, and socioeconomic-political factors (Olderman *et al.*, 1990). It is worth highlighting that over-pumping together with water scarcity and poor farming practices are the main instigators of land degradation in most lands in Jordan. The construction of a typical map representing the degree of degradation of lands resulting from these effects in a specific area could help in the assessment of the dimensions of deterioration, monitoring and the development of protection plans. Maps of land degradation provide information about degrees of degradation and can be used to assess the risks and consequences associated with it. Generally, the evolution of these manifestations can be monitored by using geoinformatics through a process known as Condition Monitoring (CM).

Several techniques have been developed to map degradation. Degradation modeling attempts to characterize the evolution of degradation signals. There is a significant number of research works that have focused on degradation models (Lu and Meeke, 1993; Padgett and Tomlinson, 2004; Gebraeel *et al.*, 2005; Müller and Zhang, 2005; Gebraeel, 2006; Park and Padgett, 2006). Many of these models rely on a representative sample of complete degradation signals developed during the last decades and help clarify the critical elements of degradation and how to assess them at different spatial and temporal scales (Trisorio-Liuzzi and Hamdy, 2002; Brandt *et al.*, 2003).

The main objective of this project is to develop a model that helps identify locations suffering from land degradation by studying the merits of the deterioration in the study area and by taking advantage of the platform for other models that have been developed. Environmental and anthropogenic factors are mapped and incorporated into the model. These parameters are extracted from known databases, field surveys, laboratory work and field interviews. The method employed is validated in order to test the reliability of the maps against data collected from other sources.

By way of criteria, the model should be simple, strong, easy to deal with during the interpretation of results, and broadly applicable. It should be possible to assess the environmental situation and degree of deterioration based on the parameters selected. Within the model, it should be feasible to modify temporal and spatial variables; it should be scalable and dispensed as required by the search and should enable the addition of transactions. The model developed was named the Land-Degradation Degree

Index (LDDI), a measure that reflects the degree of degradation in a given region. The LDDI can be calculated by using the following formula **2.3**:

$$\textbf{LDD Index} = \sum (Srxw + Vrxw + Crxw + Mrxw) \qquad 2.3$$

Where:
r: rating for the area being evaluated.
w: importance weight for the parameter.
SI: Soil Indicator.
VI: Vegetation Indicator.
CI: Climate Indicator.
MI: Management Indicator.

Each indicator contains a set of parameters, the total being fourteen. Considering the causes and impacts of land degradation, there is a growing need to develop methods for the estimation and monitoring of the phenomenon.

The most suitable combination of techniques for monitoring, evaluation of change, study of phenomena, and investigation of natural processes is remote sensing together with GIS. Remote sensing provided the necessary data for mapping land use, while spatial and modeling approaches to investigate the impact degradation on the quality of lands relied on GIS. Remote sensing data was used to estimate the salinization degree on agricultural land. The LDD Index was developed for the following reasons:

1. The availability of data: the available data for this project was not sufficient to use any mathematical or statistical model recommended in the literature. Such methods require large amounts of data collected over a long period of time. This model was developed in order to keep pace with the availability of data and to serve this field of studies.

2. The use of remote sensing: an approach known well in many applications and widely employed in works on change detection.

3. The use of GIS: the model is easily implemented entirely in a GIS environment due to the existence of the overlay procedure and the arithmetic operation functions in various GIS software, which facilitates its implementation.

4. Application: the model serves the requirements of the study; it applies arithmetic operations for monitoring the sources of deterioration, for instance climate, soil, vegetation, and management, and the magnitude of their roles through parameters that have well-known weights based on reliable and consistent criteria.

Chapter 3

Study Area and Materials

2.1. Introduction

This chapter describes the study area from which data was collected. Socio-economic characteristics like agricultural activities and inhabitants as well as the physical characteristics such as climate, topography, soils, geology, hydrology, and groundwater are being described in detail. The chapter also includes an explanation of the criteria for the sources and datasets used.

3.2. The Research Site

The research site is located in northeastern Jordan. It was chosen for investigation of land degradation by using geoinformatics methods and the exploration of the quality of groundwater and change detection of land use and land cover respectively by using statistical and spatial approaches.

3.2.1. Selection Criteria

The reasons for the selection of this test area include the following:

- ✓ The region is located in the Yarmouk basin, an important source of water in Jordan exposed to the problems of over-pumping The region is a rich agricultural zone, known for its export of vegetables.
- ✓ Random spread of agriculture in the area, and the significant change of agriculture locations.
- ✓ Availability of remote sensing and GIS data for the area.
- ✓ For comparison purposes with previous studies in the region were a motivation to pursue research in the same area

3.2.2. Study-Area Location

The study area is located in the northeastern part of the Yarmouk basin of Jordan as shown in **Figure 3.1**. 69.29% of the total study area lies within Al-Mafraq administration and 30.71% within Irbid administration. In entirety, the area under study spreads across about 256 km^2.

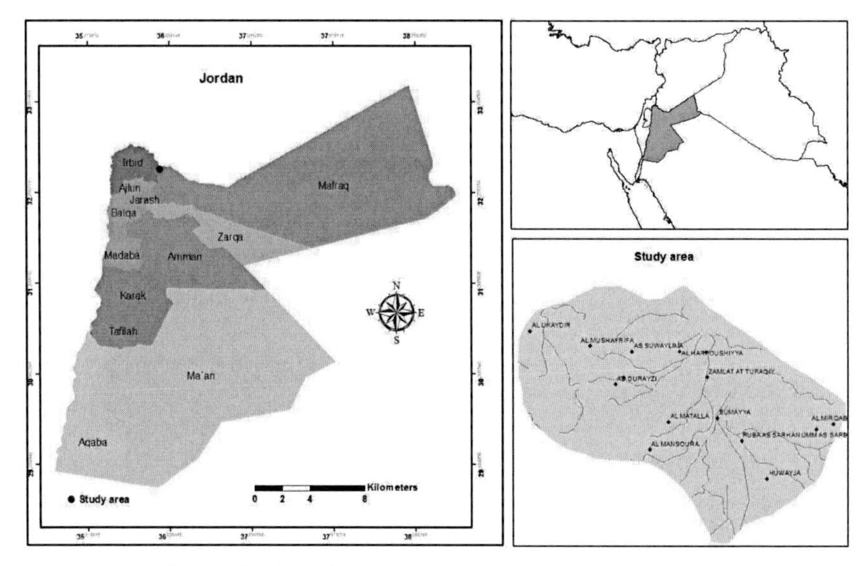

Figure 3.1: Location of study area (source: RJGC 1995).

3.2.3. Climate

Mediterranean climate prevails in the study area. It is arid to semi-arid and generally characterized by its hot, dry summers and cold winters (Salameh, *et al.*, 1997). Most of the study area falls within the western Badia region which is defined as a transition zone between the environment of the Jordan Valley and the arid interior desert areas of eastern Jordan.

This area includes all lands that receive an annual rainfall of 50 to 200 mm. **Figure 3.2** illustrates the spatial distribution of annual precipitation in the study area; the average varies between 100 and 150 mm in the north-east and 150-200 mm in the south-west. The climate is further characterized by seasonal contrasts in temperature and high variations in rainfall within and between years (AARDO, 2007). This section will describe the following climatological parameters in the study area: (a) rainfall, (b) evaporation, and (c) temperature.

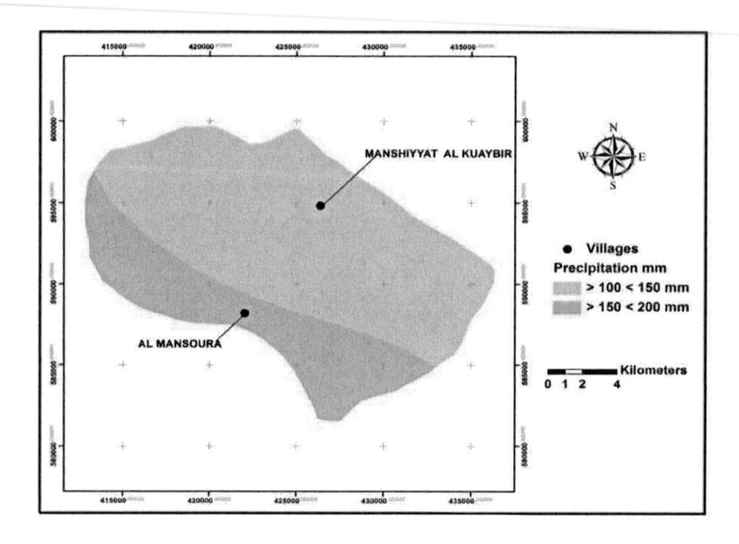

Figure 3.2: Precipitation in the study area (source: Jordan Ministry of Water and Irrigation).

The annual rate of water loss through evaporation is 92.5% (WB, 2011); climatic conditions in Jordan add to this impact in amount and distribution of precipitation and strongly affect the potential for evaporation. The estimated evapotranspiration can be fifty times greater than the mean annual rainfall (Al-Ansari and Baban, 2001), averaging 1,500 mm to 2,000 mm per year (Allison *et al.*, 2000). Temperatures in the study area vary by season. In summer, mean annual maximum temperatures reach 34 °C to 37 °C in August and occasionally exceed 42 °C. In winter, the temperature rarely falls below freezing, with annual minimum temperatures between 2 °C to 9 °C (Jordan Meteorology Department).

3.2.4. Topography

Topography varies within the study area. The lowest land, 540 m A.S.L., is near the Syrian border and hosts agricultural activities and urban settlements, while the south-western part of the study area is the highest with altitudes up to 790 m A.S.L. **Figure 3.6** shows the 10m-interval contour lines over the study area.

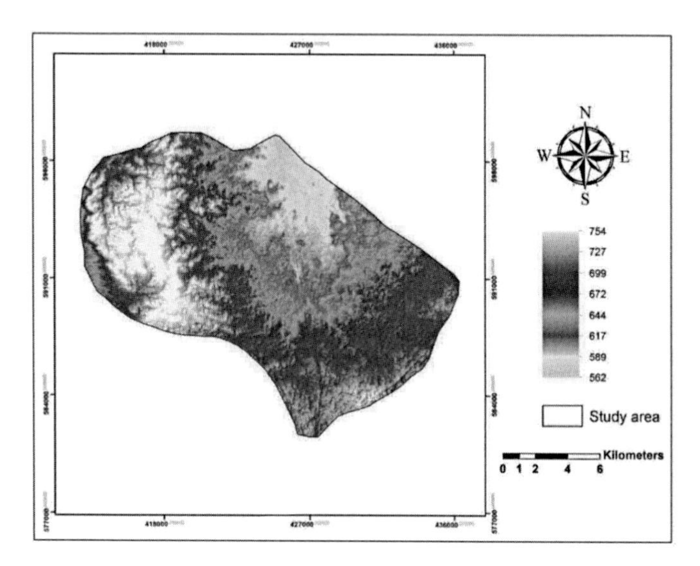

Figure 3.3: Topography of the study area (source: SRTM/X-SAR in 2000).

3.2.5. Geology

Surface Geology

In general, Jordan resides on the northwestern flank of the Nubo-Arabian shield along the eastern flank of the Jordan - Dead Sea transform and the southern shore of the ancient Tethys Ocean (Bentor, 1986). Jordanian geology has been influenced by many activities. The study area lies within the end of the basalt plateau parties. **Figure 3.4** illustrates the surface geology in the study area.

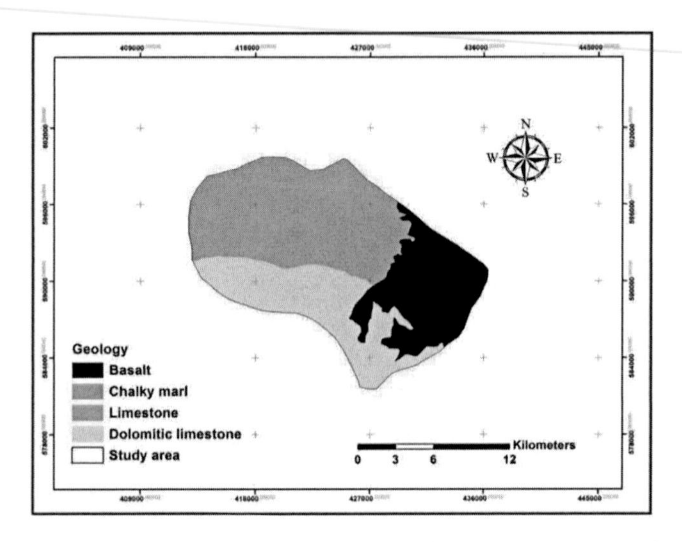

Figure 3.4: The geological map of the study area (Natural Resource Authority, 1994).

Sedimentary rocks are dominated by carbonates (limestone, dolomite, marl, and chalk). The exposed bedrock ranges from the Turonian (Upper Cretaceous) to the Pleistocene (Quaternary) and features a carbonate platform with shallow, tidal-lagoonal depositional marine environments. The platform was later drowned, and shallow to moderately pelagic facies – predominantly of chalk – were deposited from the Santonian to the Upper Eocene. A shallowing phase of the sea took place from the Upper Eocene to the Oligo-Miocene and resulted in the deposition of limestone. Regression of the sea in the Pliocene resulted in lacustrine sedimentation. From the Pleistocene to recent times, lacustrine sedimentation continued and was interrupted by volcanic activities and soil-formation processes (Moh'd, 2000).

3.2.6. Soil

Several factors still influence soil formation, types, and properties in Jordan, such as parent material, topography, and climate. The ground in the study area includes soils developed under a Torric (Aridic) Moisture Regime. Units of the Soil Taxonomy Order (Aridisols) cover about 60% of the total area of Jordan which also features the aridic soil-moisture regime (Al-Qudah *et al.*, 2001). **Figure 3.5** illustrates the soil texture in the study area.

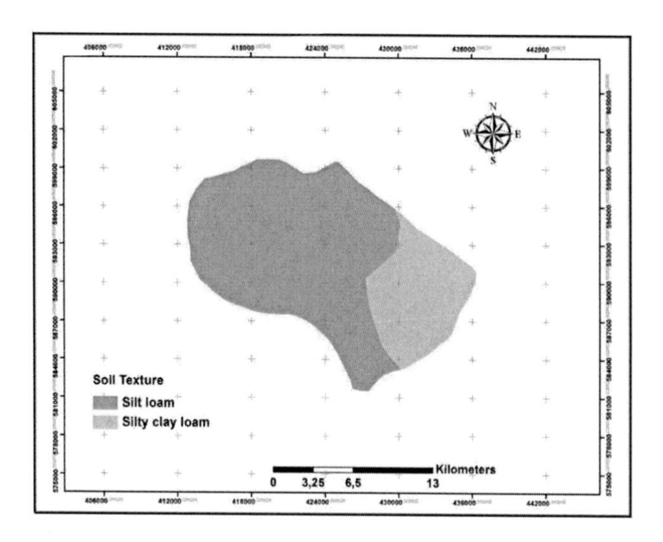

Figure 3.5: The soil texture in the study area (source: National Soil Map and Land-Use Project, The Soils of Jordan, Ministry of Agriculture, 1994).

3.2.7. Hydrology

The study area is part of the Yarmouk basin in the northern part of Jordan (**Figure 3.6**). The Yarmouk river discharges flood and base flows water as well as the groundwater of the catchments (Ta'any *et al.*, 2007). The northeast of Jordan is a key zone on the country's hydrological map (**Figure 3.1**), neighboring the highest uplands in the region east of the Jordan Rift Valley, the Ajlun mountains of the Golan Heights standing at 1,200 m above sea level. These areas receive high precipitation. They delineate the northern boundary of the study area where the Yarmouk River flows at the borders between Syria and Jordan. The Jordanian and Syrian territories are catchment areas of the Yarmouk River which originates in Jabel Al-Arab in Syria. The average discharge of this river is about 400 million cubic meters per year (Salameh and Bannayan, 1993). The total catchment area of the river measures 6,790 km^2, out of which 1,160 km^2 lie within Jordan upstream of Adasiya and the rest within Syria and in the Jordan river area downstream of Adasiya . The catchment area of the Yarmouk River is agrarian with small industries located in the main towns in Jordan and Syria.

3.2.8. Hydrogeology

Groundwater can generally be of recent or old recharge, which, for its extraction, is a vital issue of sustainability and essential for sustainable development, the amount of available groundwater in a certain area, and its renewability.

Yarmouk Basin

The 1,426 km^2 Yarmouk basin occupies the northern part of the Jordan highland and plateau (**Figure 3.6**). The quality of groundwater is variable dissolved solids concentrations range from 300 to 500 mg/L where aquifers are under water table conditions from 550 to 1,660 mg/L where aquifers are confined. Generally, the Yarmouk basin contains three distinctive drainage patterns. The northern and northwestern parts of the basin consist mainly of URC formation exposures. Areas in the eastern parts of BS (Basalt) formation exposures are dominated by a NW drainage trend. The second one, a NE trend, dominates areas in the southwest that consist mainly of Ajlun group rocks. The third and central part of the basin is dominated by ASL and MCM formations.

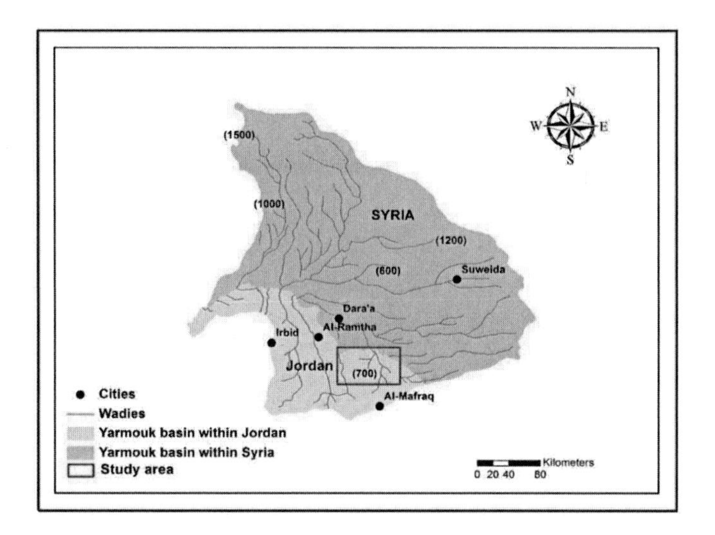

Figure 3.6: Catchment area of the Yarmouk river, 700 m ASL (Salameh and Bannayan, 1993).

3.2.9. History of Land Use

Agricultural development and water policy in North Eastern Jordan has been the subject of direct government intervention since the 1960s particularly after settling nomads. As a result, the agricultural base of the region has shifted from livestock to a cultivation/livestock mix. Irrigated agriculture in the study area started in the early 1990s after a government decision to allow wells to be dug in order to start cultivating the land in areas close to the Syrian border (Kirk, 1998). Irrigation was promoted by providing government loans for drilling private wells with the objective of cultivating more land. As more people settled, land cleared by earlier generations has been utilized and the new generation of capitalist farmers have expanded eastwards by clearing the basalt regions of the surface stone layers.

Most of the farmers grow vegetables and forage crops. Fruit trees are grown but are less abundant because of the lower short-term return they give to the farmer. The vegetable farmers tend to move their farms from one location to another around wells every year. This means that farmers rarely cultivate the land for two consecutive years in order to maximize crop production. According to references, soil salinization is the major reason for the abandonment of old farm sites.

Causes of Concern

As the human population grows in this region, more and more people are seeking new land for agriculture and other uses. As a result, agricultural land loss is a major threat. Unsustainable exploitation of cultivated land and over-pumping of water, salinization, degradation in addition to overgrazing, and encroachment all put pressure on this ecoregion's agricultural land. In some areas of Jordan, up to 50% of agricultural land has already been lost over the past twenty years.

3.3. Materials

In order to map and detect changes in land cover as well as to investigate land-cover changes, three kinds of datasets were analyzed: satellite images, GIS data (topographic and previous land-use/cover maps) and fieldwork, surveys, measurements, observations (questionnaires), and secondary data.

3.3.1. Remote-Sensing Data (Satellite Data)

Through this project on one of multispectral satellite data were relied that were available during the period 1990-2010. Selection of the multispectral dataset was principally based on the availability of landscapes and views required within a specified period for the purposes of the study.

After examining all Landsat scenes for the NE Jordan region in the United States Geological Survey (USGS) and other resource archives, several Landsat images (thematic mapper) were selected for several years in order to pursue the objectives of mapping the present state of the vegetation and of detecting change as well as to investigate the degradation of land. **Table 3.1** shows some information about images that were obtained from the USGS.

Table 3.1: Characteristics of TM data.

Band No.	Wavelength Interval (μm)	Spectral Response	Resolution (m)
1	0.45 - 0.52	Blue-Green	30
2	0.52 - 0.60	Green	30
3	0.63 - 0.69	Red	30
4	0.76 - 0.90	Near-IR	30
5	1.55 - 1.75	Mid-IR	30
6	10.40 - 12.50	Thermal-IR	30
7	2.08 - 2.35	Mid-IR	30

3.3.2. GIS Data

GIS data comprises the historical land-use maps and the Shuttle Radar Topographic Mission (SRTM) Digital Elevation Model (DEM) – 25 m resolution as well as a host of other data appearing in **Table 3.2** in addition to the sources.

Table 3.2: GIS data and sources.

Data Type	Year	Data Format	Data Source
Hydrology (wadis)	1994	Digital	MWI
Hydrology (basins)	1994	Digital	
Land-use map	2012	Digital (Digitized)	Google Earth (Geo eye-1)
Temperature map	2012	Digital	JMT
Rainfall data	2012	Digital	
Geology	1994	Digital	NRA, Fieldwork
Soil map	1994	Digital	

(NRA): Natural Resource Authority

(MoA): Ministry of Agriculture

(MWI): Ministry of Water and Irrigation

(JMD): Jordan Meteorological Department

3.3.3. Secondary Data Collection and Analysis

I. Rainfall Data

Rainfall Data was used to modeling groundwater vulnerability as well as in the land degradation management. Data were collected from the *Jordan Metrological Department* (JMD); they were available for the Al-Mafraq station and some neighboring areas **(Appendix B-I).**

II. Geology

The Natural Resources Authority of Jordan (NRA) has produced geological and mineral occurrence maps during past three decades, The Natural Resources Authority of Jordan (NRA) was produced a Geological and Mineral occurrence maps at different scales over the past three decades. The geological map of the northeast of Jordan includes (a) lithological structures, (b) surface geology, and (c) pervading fault lines, these maps were obtain on digital format. The surface geology of the study area is presented in **Chapter 3, Figure 3.4.**

III. Topography

Topographical maps in digital format were prepared by the Shuttle Radar Topography Mission (SRTM/X-SAR) in 2000. The methodology for producing the topographical maps at SRTM was based on using two single-pass SAR interferometers (C- and X-band) to produce the most consistent digital elevation model (DEM) with a horizontal resolution of 25 m. The topographic map of the study area is presented in **Chapter 3, Figure 3.3.**

IV. Soil

Jordan Ministry of Agriculture (MoA) carried out a project between 1989 and 1993 to survey the soil in Jordan as a national project in order to produce a soil map in collaboration with outside organizations; the soil map was used within this research in digital format, as well all data collected about soil from other sources. Soil mapping was produced based on remote-sensing data Landsat imagery and aerial photographs with taking into account field observations as well as observation of soil sectors within estimated distances (Jordan Ministry of Agriculture, 1994). The soil map of the study area with different data of soil classes is presented in **Chapter 3, Figure 3.5.V.**

Land Use

The available land-use map was digitized by using Google Earth (Geo eye-1) satellite imagery that represents the current status of land use in the study area. The map that was digitized was taken by a high-resolution earth observation satellite which had been launched in September 2008. The imagery of the study was acquired on 11th May 2012.. The spatial resolution used is very high with pixel sizes of 0.41 m. The production of the land-use map will be discussed in detail in **Chapter 5, Section 5.2.4.**

3.3.4. Fieldwork Data

Terrestrial surveys provided ancillary data to geo-reference satellite imagery and to facilitate their interpretation. It provided subsidiary information to the nature of the study site.

The ancillary data (information) acquired during field surveys include:

- ✓ Ground truthing;
- ✓ Identification of land-use and cover classes;
- ✓ Records of training and accuracy-assessment data;
- ✓ Examination of areas that have experienced conversion of land use/land cover;
- ✓ Collection of samplings for soil and water analyses.

The procedure for the acquisition of this ancillary information is elaborated in the methodology.

3.3.5. Software and Hardware

The following soft- and hardware were employed in at least one phase of the study:

Erdas Imagine 10: image pre-processing and thematic information extraction (pixel-based classification); digital change detection.

ArcMap 10.2: spatial analysis and modeling – model builder.

GPS 72H: acquisition of ground-truth and accuracy-assessment data.

Excel, SPSS: feature selection and multiple models.

R studio: Analysis and statistical relationships.

Methodology

4.1. Introduction

The methodology chapter contains sections about remote-sensing processing, validation of the data that was collected, and image classification, while also explaining the methodology for change detection that appeared in land use in the study area twenty years ago; another sections includes and describes the methodology for fieldwork and labs analyses of groundwater hydrochemistry for 2012 and 2013 – as well as soil-chemistry data (geochemistry), and there is some descriptive statistics, as well as a section about geoinformatics data processes of groundwater vulnerability and land degradation methods with integration.

4.2 Methods

The methods implemented in this study can be broadly divided into three subsections, cf. **Figure 4.1**:

- Image interpretation and change detection (remote sensing) where the land cover change from 1990 to 2010 was calculated to use the results in the some equations and to support interpretations of images.
- Field survey and laboratory analysis (socio-economic and soil analysis) where the samplings of soil and water were collected in order to carry chemical and physical analyses that will be used for the explanation of results and to assess the DARSTIC index and a development of the land-degradation model.
- Building the degradation models (modeling); this section contains the factors and results of modeling and the methods used to validate the model.

4.2. Remote-Sensing Processing

In order to map and to detect changes in land cover, three types of data sets were analyzed: satellite images, DEM Map 25 m from USGS and a pre-existing land-use/cover map on a scale of 1:250,000 by the Ministry of Agriculture, as well as a mapping of land use by using the Google Earth.

4.2.1. Image Pre-processing

The pre-processing of satellite images prior to actual image analysis (establishment of thematic maps and change detection) is essential; its unique goals are the establishment of a more direct linkage between the data and biophysical phenomena, the removal of data-acquisition errors and image noise and remote-sensing image pre-processing includes the following steps: geometric correction, atmospheric correction, noise removal, and image enhancement. Atmospheric correction and noise removal were not implemented in this study because data relating to different atmospheric parameters were not available during image acquisition.

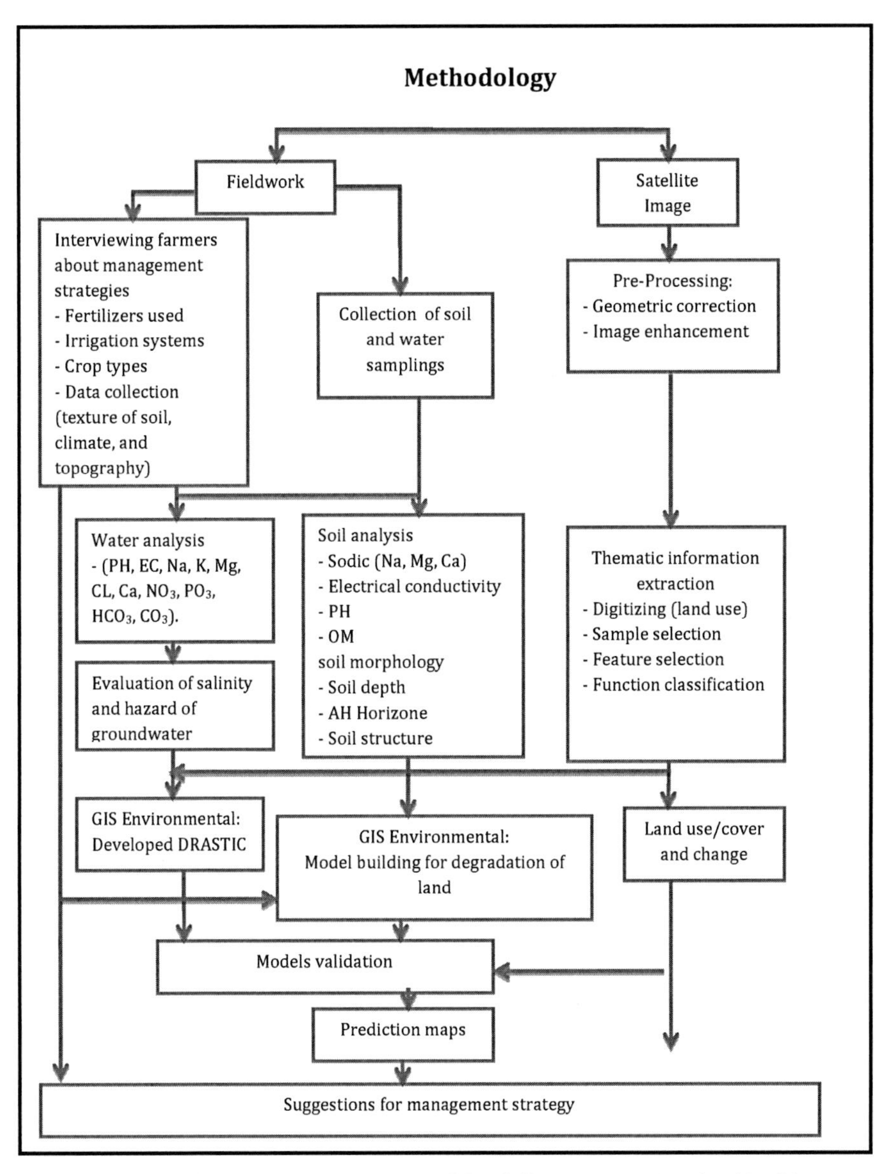

Figure 4.1: Schematic representation of the different stages involved in this research project.

Distortions and noise occur in remote-sensing images during their acquisition. This is principally due to: variations in atmospheric conditions, altitude, velocity of the sensor platform, platform perturbations, scan speed, and the swath of the sensors' field of view, earth curvature, and relief displacement. It is therefore necessary to correct these errors before proceeding with the extraction of thematic information from the satellite images. Apart from correcting the effects of distortion, pre-processing techniques also increase the visual distinction between features.

4.2.2. Geometric Correction

Before the classification process, the Landsat images were geometrically corrected. The methods to do so can be divided into two steps depending on the types of errors. The systemic errors were corrected at the receiving station. The random distortion was corrected by selecting twenty ground control points (GCPs) from topographic maps, which could be easily localized on the satellite images. The polynomial function was then used to correct the coordinates of the image positions and to form an undisturbed output grid. After this, each cell in this new grid was assigned a grey level according to the corresponding pixel in the original image in a process referred to as 'nearest neighbor resampling'. Because the cells in the original image and the ones in the new grid do not overlap, the Digital Number (DN) values cannot be assigned by simply overlaying the two; instead, this is done through interpolation methods. In the nearest neighbor algorithm, the pixel is assigned the DN value of the closest pixel in the original image. This method keeps the original image with spectral information and (in the cases of half pixels) makes the images slightly disjointed because of the mismatch. The Landsat data were georeferenced to UTM zone 36 N, with the JTM (Jordan Transverse Mercator) being based on the International Hayford Ellipsoid 1924.

4.2.3. Geo-referencing for the Satellite Data

Satellite images have a Universal Transverse Mercator (UTM) coordinate system which was obtained from the United States Geological Survey (USGS), and this study carried out the process of projecting these images by using the coordinates of fixed reference points. **Figure 4.2** shows such items as roads and dams in the study area. We used shapefile maps in the re-projection process to the Jordan Transverse Mercator (JTM) coordinates with ArcGIS and ERDAS. The satellite images were geo-referenced to JTM by using the (re-projection) command. This command transforms the raster dataset from one projection to another, covering the same area.

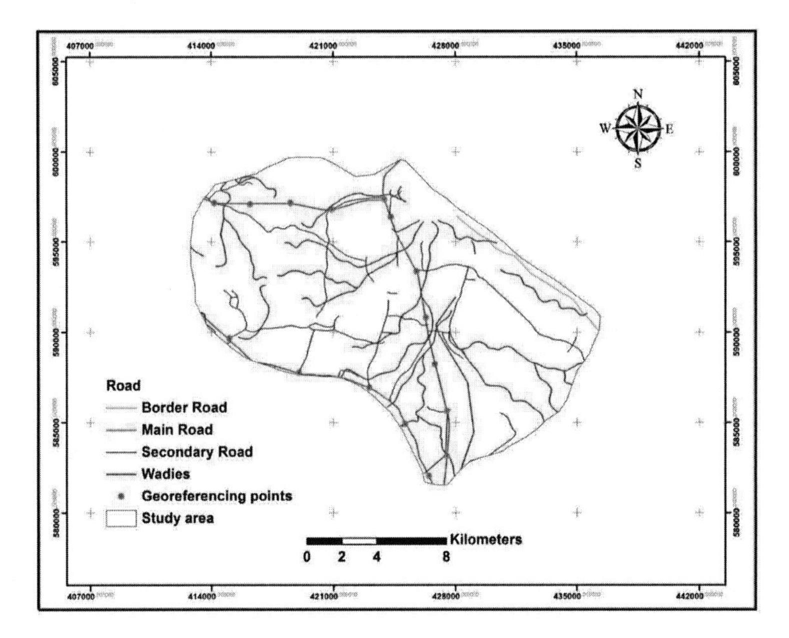

Figure 4.2: Points used in the geo-referencing process.

4.2.4. Monitoring of Residential Areas

The residential areas (villages, towns, and large industrial facilities) were mapped manually (digitizing) by using Google-Earth (**Figure 4.3**). The vector file that contained these polygons was exported to ArcView GIS where their file was attributed and data that was recorded from field surveys or obtained from outsources was updated.

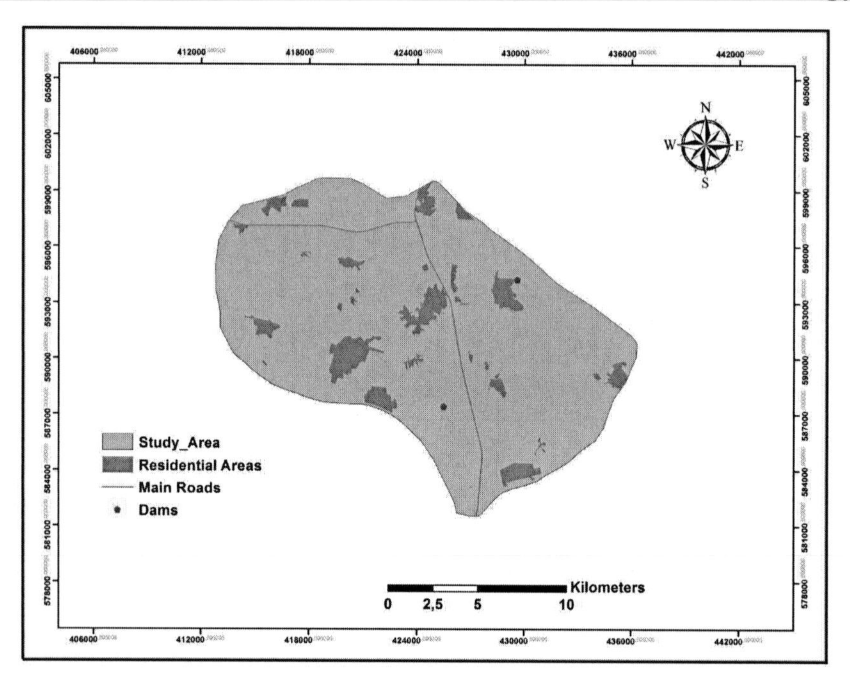

Figure 4.3: Residential areas map based on satellite image 2012.

4.2.5. Spatial Resolution

The spatial resolution for the satellite images should be considered before conducting data analysis in order to overcome the discrimination problem, which means providing the ability of distinguishing farmed areas from the surrounding landscape features through pixel sizes that exist in the available satellite images. In this research, the spatial resolution issue was dealt with by using Google Earth which has a high spatial resolution of up to 0.5 m that could distinguish the landscape in the study area and can cover all areas by digitizing these images in order to map the farming activities and land use in the study area. The Landsat satellite images have a spatial resolution of up to 30 m, with a pixel size of 900 m^2, which provides the ability to use the Landsat TM with 30 m resolution to map the farming activities and to investigate the change of land use/ cover with existing reference points in the study area.

4.2.6. Processing Bands

The application of image-enhancement techniques served to improve the interpretation of the image by increasing the apparent distinction between the land use and land cover classes in the scene. The following image-enhancement techniques were used in the transformation of the satellite data: several indices like the degradation index and salinity index.

In order to use the digitized command, it was necessary to define the best combination of bands that could be used to extract the required information. Since this research aimed at investigating agricultural land-use changes, bands TM3, TM4 and TM5 were selected. **Table 4.1** summarizes the practical use of these bands in mapping agricultural land use. Bands 3, 4 and 5 were processed by using the digitized command, which produced components that could be reliable (e.g. bands 1 and 4) as shown in **Figures 4.4 and 4.5.**

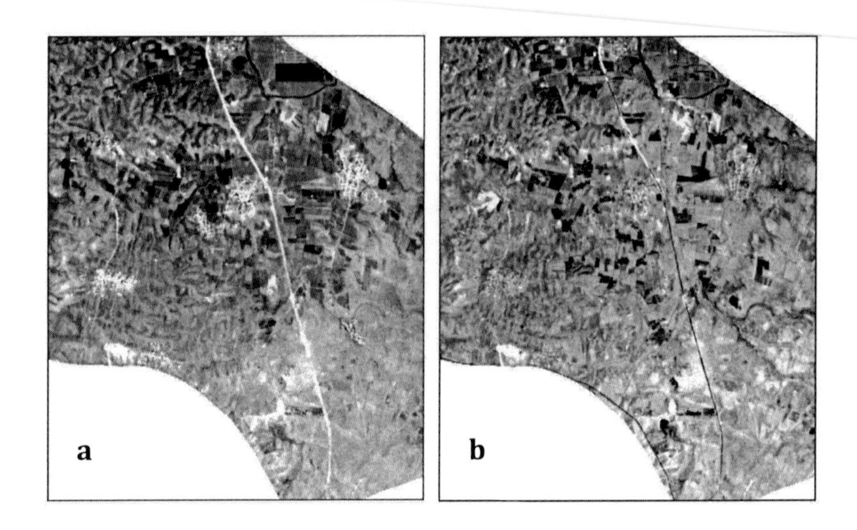

Figure 4.4: Bands of Landsat TM as of 1990: (a) 1st component, (b) 4th component.

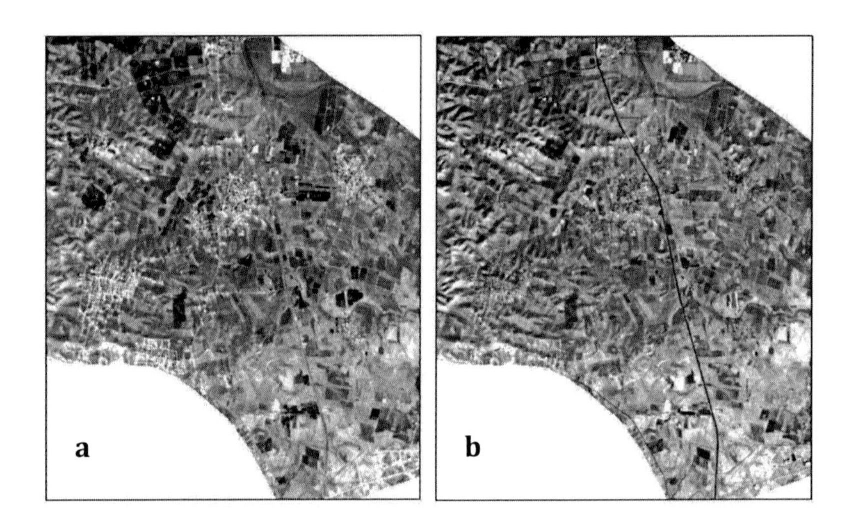

Figure 4.5: Bands of Landsat TM as of 2010: (a) 1st component, (b) 4th component.

Table 4.1: Landsat TM band applications (after Campbell, 1996; Wilkie and Fin, 1996; Barrett and Curtis, 1999).

Bands	Descriptions
5	Moisture content in soil and vegetation; vegetation type discrimination and plant vigor
4	Plant biomass studies; soil-crop and land-water boundary discrimination
3	Discrimination of vegetation types

A. Degradation Index

The degradation index defines the degree or stage of degradation that can be secondarily used as a measure of degradation that reflects the reclamation degree, where is necessary to assign the degraded land that appeared in the analysis sodic in the soil sampling, the degradation index is computed by following the next formula as recommended (Zhao and Meng, 2010).

$$\text{Degradation index} = 255-(B2+B3)/255+(B2+B3) \qquad \mathbf{4.1}$$

where

$B2$ = blue band (band 2 in TM data)

$B3$ = red band (band 3 in TM data)

B. Salinity Index

Soil salinization is more serious and has become the main form of land degradation; remote-sensing technology was used to detect soil salinization as a degree/stage of degradation by using the salinity index as recommended (Al Khaier, 2003):

$$\text{Salinity index} = (B4-B5)/(B4+B5) \qquad \mathbf{4.2}$$

where

$B4$ = near-infrared band (band 4 in TM data)

$B5$ = middle-infrared band (band 5 in TM data)

4.3. Data Validation

4.3.1. Royal Jordan Geographic Center Data

An investigation of these data and maps (of roads, dams, towns, villages, wadis, and farms) was used for ground-reference points by GPS as well as satellite images with high resolution, data and maps were validated through return to the workers in RJGC who produced such maps through methodology used and were obtained with full required details about mapping data.

An error could occur during the mapping process due to the use of GPS with ground-reference points. The locations of these points are illustrated in **Table 4.2**, which shows that the shift between both datasets was less than 0.25 (*Google Earth*). This could be considered to be an acceptable difference because the satellite used in the process of mapping has a high resolution of up to 0.5 * 0.5 m. Besides that, the portable GPS that was used had a high accuracy because it depends on seven satellites with an error rate of less than 15 m, while the usual rate can be up to 95%; it had been recommended by Al-Adamat (2002) for data validation through fieldwork during this research. On the other side, it was found that the surface geology map matched the ground in the inspected area and what appears on Landsat TM with a resolution of 30 * 30 m. This means that the published geological map is valid and could be used as a reliable source of data for this research programme.

4.3.2. Quantification of the Change in Land Use

The results suggest some differences in farm size between the figures provided by the farmers and the actual farm size as derived from the satellite image for the year 2012, where 12 farms were used from all farms in order to quantify the change in land use. **Table 4.2** shows the actual farm size as calculated from the satellite images, together with what has already been mentioned about the size of farms. The latter was also assessed in other ways such as by GPS.

Table 4.2: The farm areas as estimated through the Images Satellite (2012) together with data provided by farmers, the estimated digitizing error, and GPS.

Farm Code	Area (m^2) (Interview)	Area (m^2) (Images Satellite)	Area (m^2) GPS	Difference (m^2)*	Difference (%)	Digitizing error ($\pm m^3$)**
FFI	120600	120742.36	120742.36	-142.36	-0.1179	7.57
MAI	81300	81255.3	81255.3	44.7	0.055012	3.92
AKI	349350	349300	349300	50	0.014314	2.28
BZI	224250	224243.5	224243.5	6.5	0.002899	3.35
MKI	38850	38800.1	38800.1	49.9	0.128608	3.23
MKII	51520	51509	51509	11	0.021355	2.56
BZII	195990	195986	195986	4	0.002041	3
MAI	323490	329502	329502	-6012	-1.82457	3.85
MDI	84590	84663.5	84663.5	-73.5	-0.08681	0.74
YAI	753200	753275.5	753275.5	-75.5	-0.01002	3.39
EI	48480	48461.6	48461.6	18.4	0.037968	2.98
AJI	4800	4786.4	4786.4	13.6	0.284138	1.37
Total	2276420	2282525.26	2282525.26	-	-	-

* The results of this column refer to the result of the difference between the shaded columns.
** The digitizing error was estimated by accuracy assessment for the areas estimated by using GPS.

These differences can be traced to several possible reasons:

1. In cases where the size of farms – according to what had been recorded through farmers – was greater than the actual farms size, the difference can be traced to the fact that there is neighboring land on which plant varieties grow in the presence of moisture, resulting in an increase the farms in the area; these farms are calculated as single farms throughout the year, specifically at the time the image was captured.

2. In cases where the actual farm size was more than what had been recorded through farmers, there are several possible explanations:

A. The government took a range of decisions about issues pertaining to farms, such as the total productivity and regular monitoring by the Ministry of Agriculture and others. For that reason, farmers had worried about the questions asked in the interview and about the probability of having certain taxes imposed on them or having certain actions taken against them by the government.

B. The area of crops differs between the time the image was captured and the time the fieldwork was carried out.

C. The farmers decided to increase the area of their farms after completion of field surveys and interviews.

Several methods were used to assess the size of farms – although they were scattered around the study area, it should be noted that during the field survey it was observed that farmers moved cultivated areas in circles around the respective water source. Al-Adamat (2002) had instructed this in order to minimize the cost of pumping the water over long distances. In general, it was noted that these wells are suffering from over-pumping, which means a reverse process would threaten the surrounding areas unless handled properly.

4.4. Image Analysis

As mentioned above, the pixel-based image-analysis classification was performed by using special software. The basic processing units of pixel-based image analysis are image objects, and it was used to extract land use and cover classes in the whole study area on all scenes (Landsat TM). The process analysis can be divided into classification and accuracy assessment. Since the different segmentation levels tested in this study have already been explicitly documented in the methodology, this section will focus on presenting the results of the classification and subsequent accuracy assessment. As mentioned before, object-oriented image analysis was used to classify the whole study area.

The classification of the whole study area was important to monitor the changes in land use and land cover, i.e. to give indicators of land-use increase or decrease (especially agriculturally) in relation to the active land reclamation, as well as to see effects of the degradation processes on these lands. In the study areas, classification was used to monitor the ability of processes to separate the land types because both old and present agricultural lands have very small land parcels and different crop types in confined spaces; this is different on dry soil that is widely spread. Also, the extent of applying remote sensing and image interpretation for the land-use types in Jordan was tested in these vicinities. The pixel-based classification was performed by using Landsat (TM) images. With discriminant analysis, the bands were determined that could best separate classes in the whole study area.

A. Image Classification

Two main approaches were used to extract information from the imagery in an attempt to compare their effectiveness: technical classification and visual interpretation.

I. Technical Classification

A digital classification approach was used to map land use of the study area. The classification approach was applied for each year separately in order to automatically categorize all pixel values of an image into land-use classes. A supervised-classification approach was used in this study. Supervised classification is the process of classifying the unknown-identity pixels by using samples of known identity. These are pixels located within training areas.

The classification of the study area was done in accordance with classes. The standard nearest neighbor was selected to extract thematic classes in order to classify the main classes in the study area. For example in the agriculture land were merged trees and vegetables as farms. The quality of land-use and land-cover classification maps was assessed by using a confusion matrix which is generated either through random sampling of objects or by using test areas as reference data. This is necessary in order to ascertain the suitability of the maps for specific applications.

II. Visual Interpretation

A visual interpretation approach was used for identifying different cover types of land use based on tone, texture, shape, and relation of one area to another within the study area. The method is a low-tech approach and does not require sophisticated equipment, making it quite accessible to a wider variety of users. However, the visual interpretation approach was used to delineations and close control, and in order to detect the change between land-cover classes in the study area, the manual interpretation is key to correct classification.

B. Training-sample Selection

Training samples were used to guide the classification algorithm to assigning specific spectral values to appropriate informational classes. Spectral signatures can be developed from training areas in the image. Training-sample objects were defined in order to implement the nearest neighbor algorithm. Initially, the samples were selected based on previous analyst knowledge of the spectral reflectance of various features in the study area. Later, this information was confirmed during the field studies, as well as the acquisition of more training data, by using GPS measurements.

C. Accuracy Assessment

Classification is the process of assigning variables to discrete categories of useful information. Land-cover classification uses remotely sensed reflectance or radiances to determine the category to which a given pixel belongs. However, errors of land-cover classification occur because of soil-background differences, positional errors, land-cover mixtures (mixed pixels), or human errors.

The assessment of the accuracy was conducted in ERDAS for the whole study area. The ERDAS software assesses the accuracy based on an error matrix by using ground-truth data and (by using training samplings) are compared to the location and class of each ground-truth object (column) with the corresponding location and class in the classified objects (rows). Each column of the error matrix represents a ground-truth class, which defines the true class of the objects. The values in the column correspond to the amount of objects in that class, with this object-oriented algorithm classified to various classes.

200 random points for the whole study area in addition 200 points were set aside to accuracy assessment. These points were overlaid on the respective classified images in ArcGIS. The classes of images with the points were selected and the respective attribute tables joined. The "new" object shapefile was converted to a raster, which was then converted to a thematic layer in ERDAS Imagine.

The classification was then evaluated by using producer accuracy, user accuracy, overall accuracy, and the kappa index. **Figure 4.6** shows the steps of the accuracy assessment in this study.

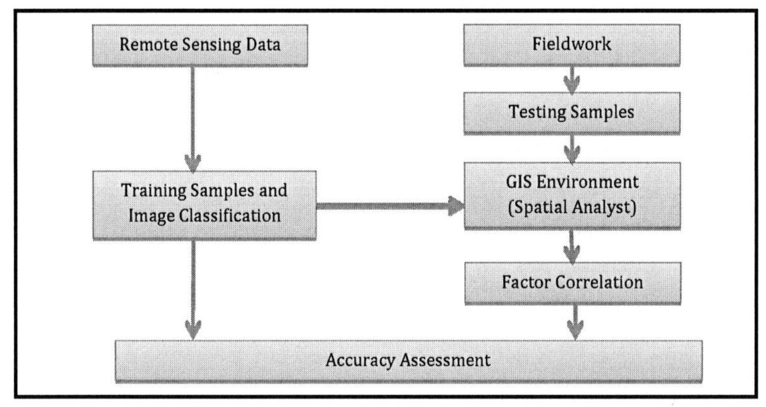

Figure 4.6: Stages of accuracy assessment.

E. Change Detection

Change detection was done through comparison results of classification after classifying the rectified images in the period (between 1990 and 2010) separately. Each date of imagery was satisfactorily segmented. The classified images were then compared and analyzed by using ERDAS Imagine and the change-detection subscript in ArcGIS to determine the change-detection matrix, and to finally construct the change maps (Huang and Hsiao, 2000).

4.5. Fieldwork & Laboratory Analysis

The fieldwork was conducted from March to May 2012 and from March to April 2013 for primary and secondary data collection. The first part was carried out in the study area where soil and water samples were collected for analyzing the parameters required for this research. The second part of the fieldwork was devoted to conducting interviews with the relevant farmers in the study area in order to obtain data on management practices. The third part was spent on secondary data and information collection from the different organizations involved in the study. They included: for soil and agriculture-related information different institutions of the Ministry of Agriculture of Jordan (Soil and Agriculture Research Centre, Statistic Research Centre), and for climate-related information the Department of Meteorology.

Information pertaining to crop management (crop-production data, fertilizer-application and soil data) and socio-economic parameters (agricultural occupations, the gross income from agriculture, the distribution of settlements, and the names of the farmers' regulars etc.) were obtained from archives (Ministry of Agriculture and Department of Statistics) as well as through interviews with experts in Jordan. Ancillary GIS data (soil maps, land-use and -cover maps) were acquired from the Royal Geographic Center (RGC). The fourth part of the fieldwork comprised collection of ground-truth information related to land use and land cover – in order to verify the land-use and -cover maps. The data collected in the fieldwork is included in four parts:

(1) Socio-economic data.
(2) Groundwater data.
(3) Soil data.
(4) Remote-sensing data.

4.6. Socio-Economic Data (Questionnaire: Interview with Farmers)

A questionnaire was designed to help in the evaluation of the field data on precision farming. Farmers were questioned about a range of management practices: cropping, tillage use of fertilizer, weed control, insect pests, yields, historical land use, etc. The questionnaire consisted of yes/no options, multiple-choice and short-answer questions. The socio-economic data includes the methods of data collection and the collected data (farm activities and fertilizer use etc.).

Methods:

First method:

Visual control (observations) was involved to make sure the data is accurate and to replace the individual's behavior, which was based on asking questions in this research project where several agricultural activities carried out in the study area were subject to modeling by geoinformatics techniques and recording (Al-Adamat, 2002). This method was used to collect data about (1) farm types and (2) fertilizer and pesticide storage.

Second method:

There are two types of questionnaire surveys: (a) descriptive processes supposed to find descriptive facts and to collect accurate information on a particular phenomenon; and (b) analytical survey processes that explain and clarify a phenomenon (Oppenheim, 1992).

In this research study, in order to describe the agricultural activities that were distributed across the study area, it was necessary to use a descriptive questionnaire survey rather than an analytical one, whereby some interviews were carried out with people involved in agriculture in order to collect the required data about farm practices and methods. Descriptive questionnaire surveys included questions that were either objective or optional; explicit questions based on the principle of selecting one answer among alternative ones were asked to the people involved in agriculture. Dialogic questions were open, took a long time to be answered, and were complicated in terms of the commentary.

The reasons leading to the use of dialogic questions were as follows:

(a) What interviewees say and what they think can be noted directly, and according to Oppenheim (1992) and Al-Adamat (2002), the main advantage of the open question is the flexibility it gives to the respondents.

(b) The number of farms visited in this survey was 43 farms. The project was got approval to sampling from 22 farm and interviews owners of these farms, which made it easy to carry out of fieldwork by using dialogic questions.

(c) Direct knowledge about any details gained from participants involved could be referred to during the interview. The questionnaire in **Appendix A-I** was divided into three major sections:

1. Well data included location and depth to groundwater surface.
2. Farm data included area, crop type, fertilizer and pesticide types as well as application rates.

3. The agricultural practices and provided extension services to farmers by organizations were useful for the project as well as centers, and relevant ministries concerned with farms and agricultural land affairs.

Fieldwork was carried out between 4th March and 4th April 2012 in order to collect data about wells and the farming activities in the study area. 22 farmers were interviewed. The interviews took place on their farms between 11:00 a.m. and 3:00 p.m. by appointment throughout the week, except on Fridays and Saturdays. Timing the interview was an important factor as to make sure that farmers were available at this time of day.

4.7. Farming Activities

43 farms were visited, ranging from productive to unproductive ones within the study area at the time of fieldwork. 22 farms were consulted for the collection of samples that shown in the **Figure 4.7**. There were a number of difficulties that prevented the completion of some surveys in this study area.

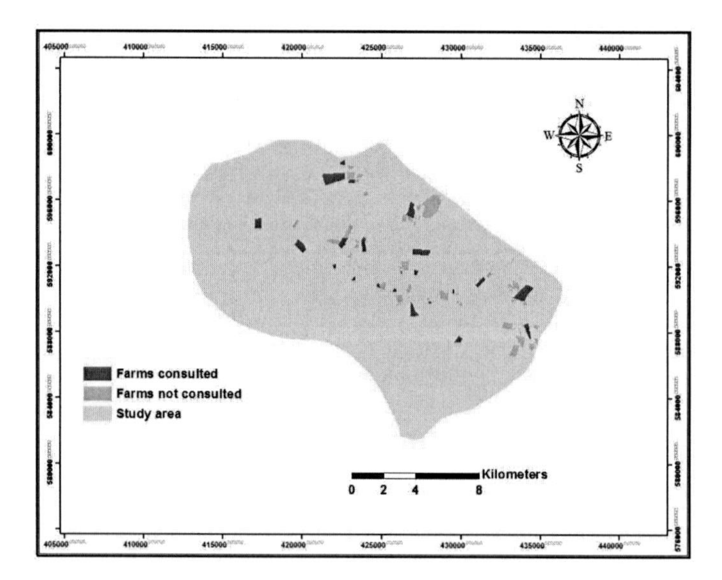

Figure 4.7: The consulted and not consulted farms in the collection of samples.

Some survey questionnaires were distributed to unclassified farms, which mean no information existed about them or there was a lack of knowledge about their owners, while other farms refused to collaborate in the project, did not contain wells, or were located in wastewater areas. In addition, 21 sites were visited that represented abandoned farms. The results of the field visits showed that agricultural practices of 14 farms in the study area involved seasonal crops (vegetables), and 8 of these owned practiced both types of agricultural production (trees and crops).

Furthermore, it was noted that farmers stored fertilizers and pesticides close to the wells. Some of them (16 farms) had a small store in which these products were kept, but others (6 farms) kept them outside, only a few meters away from the wellhead. This could be due to the fact that all wells had a guard or operator who might also be employed to protect these products.

Farmers were asked to provide information about the size of their farms and which types of crops they had been growing in the farming season of the year 2012. The data collected from farmers included the types (organic and inorganic) and amounts of fertilizers and pesticides used, and additional information was obtained about the respective fertilizers or pesticides - such as name and order - in order to locate their import source. It was difficult for farmers to estimate how much they used of each type (fertilizer or pesticides), for several reasons:

1. Because these pesticides depended on the existence of a specific disease, the farm area, and how extensive the infection was.
2. Lack of knowledge on the part of farm workers about the amount of fertilizer to use.

There are more details with regard to fertilizer types used as shown in **Table 4.3**. According to the farmers, in conditions where infection breaks out in one area of the farm, the organic fertilizers used increase the quality of soil in order to increase productivity. **Table 4.4** shows their chemical components; as for pesticides, it suffices to mention the most commonly used types: VERTIMEC, VAPCOMIC, BAYFIDAN, AGRINATE, DEFOS, NEORN, LANNATE, KARATE, DECIS, PROPARGITE, TRAKER, TOPAS.

Table 4.3: Fertilizer which farmers used to improve of soil quality.

Fertilizer name	Rate of fertilizer use for the region
Urea, Map, Chicken Manure, Sheep Manure, Singral	High
Nebras, Iron, Ammoniac, Fertal, KNO3, Mannalina, Twenties, Peters, Amkolon Twenty	Medium
Magnesium Sulphate, Sekrteen, Potash 36,Fertland, Mixed Manure, MKB, Green Power, Netroleaf, Potash 36, Cuzan, Kabreet, Liquid Phosphor, Fardous, Ammonium Nitrate, Calcium Nitrate, Nirosol, Zinc Sulphate, Netrofoska, Total Grow, Amcopeter	Low

Table 4.4: Chemical components of these fertilizers.

Formula	% of Total weight			Usage rate
	Minimum	Maximum	Mean	
N	10	46	23.5	High
P_2O_4	4	61	21.3	
K_2O	4	46	17.5	
NH_4	14	20	19.2	Medium
NO_3	10	13	11.5	
Zn	-	-	36	Low
Mg	-	-	16	
Fe	-	-	6	
S	-	-	9	
Ca	-	-	16.5	

The questions asked in the last part of the questionnaire were about the extension services provided to farmers and the reason for changing the location of agriculture annually, as well as the agricultural practices used regarding fertilizers and pesticides. The great majority of farmers (more than 40%) mentioned that they depended on fertilizer and pesticide dealers to get guidelines of agricultural. This result about extension service information was also reached by Al-Adamat (2002). On the impact of fertilizers on groundwater, he mentioned in his study that most of the workers there were non-specialists and come as expatriate workers only to do a living. Only 20% of the farmers have engineering specialists as supervisors on the agricultural process and a mechanism to select the suitable quantities of fertilizers and pesticides, how to apply them, and how to deal with them. The answers that were given to some of the questions in this part of the questionnaire are summarized in the following points:

1. Providers of extension services as mentioned by farmers in 2012. The number of people who rely on fertilizer dealers is 8 (about 40%), while 6 (about 30%) rely on tradition, 4 (about 15%) on their own experience, and another 4 (about 15%) on extension engineers.
2. Extent of the spread of diseases in some agricultural farm areas from neighboring farms in 2012.
3. Two farmers (9%) complained about the spread of diseases from neighboring farms, while 20 farmers (more than 90%) did not.

These are the general results of the farming activities part, and as follows are the main reasons for annually shifting the location of cropping :

1. Through laboratory analysis, it appeared (see **Appendix B-II**) that the soils lacked organic matter needed in soil, which reflected negatively on soil degradation and salinity. As a result, the farmers were forced to use chemical fertilizers, which mean the lands not yet used have a more suitable soil quality for crops.
2. The agricultural sector lacks oversight by competent authorities.

4.8. Water and Soil Sampling

4.8.1 Groundwater Data

For most of the farmers, it was difficult to answer some questions – such as the one about the depth to the water table and the date at which cultivation started – because most of them were not the original owners of the wells. Most had either rented or bought them and the surrounding land from the original farmers. However, other data were obtained through the farmers, as well as data required from the Jordan Ministry of Water and Irrigation (**Appendix B-I**), and a hydro-chemical analysis was carried out for the data that was collected about the wells. This will be described in the next section.

The Hydrochemistry Data of Water

Water samples were collected in two stages. The first stage started in March 2012, when samples were collected from three government pumping stations (**Figure 4.8**), while others were collected from 12 private wells in March 2012. The second stage took place in March 2013, when samples were gathered from the same places as in 2012, which came as a reprocessing of the previous results. The Ministry of Water and Irrigation was contacted to get data about the water in the study area, but it turned out that the MWI did not have any records on hydro-chemical data for the private wells.

During fieldwork, it observed that there are people had removed the water taps that could be used to collect water samples in government wells (according to the water authority) due to the misuse of water by these people, while private wells had water taps that could be used to collect water samples. It was also was observed that there a distances between pumping stations and water taps, which vary from one station to another. They range from 20 m to 200 m and can sometimes measure up to 500 m or more. The samples from private wells were collected by using sterilized one-liter plastic bottles, and samples of drinking-water wells were collected before adding chlorine, the samples were transferred to a laboratory at the University of Al al-Bayt for hydro chemical analysis that were discussed in the next sections. The procedures and results of analyses were shown in **Appendix B-I**, and furthermore, **Table 4.5** includes the methods of hydro-chemical analysis.

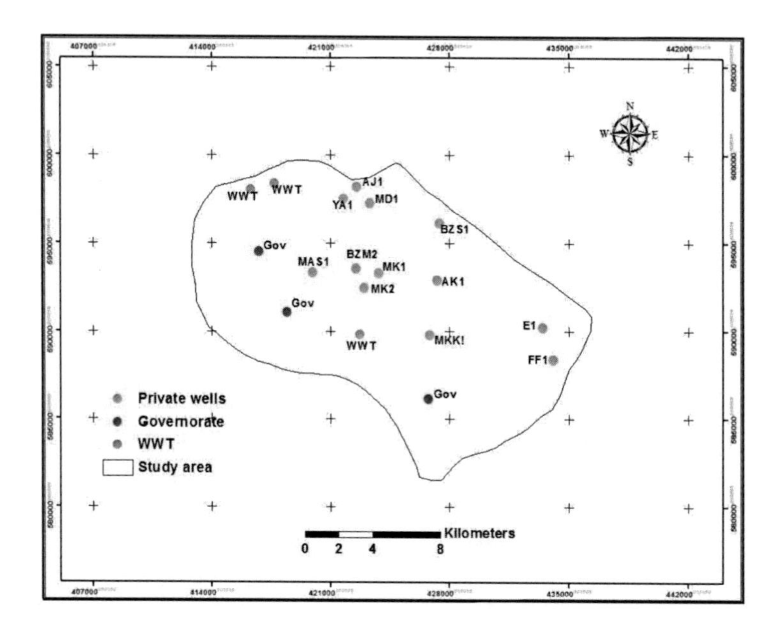

Figure 4.8: The location of private wells and governorates.

Table 4.5: Measurement methods, limits of detection and accuracy of hydro-chemical analysis.

Chemical element	Measuring method	Limit of detection	Accuracy
Na and K	Flame photometer	≤ 0.2 ppm	± 0.2 ppm
HCO_3, CO_3, Cl, Ca, and Mg	Titration	1 ppm	± 0.2 ppm
NO_3, PO_4, and SO_4	Spectrophotometer	0.001 ppm	± 0.001 ppm

In fieldwork, a GPS was used to validate data provided by the Jordan Ministry of Water and Irrigation as well as the Ministry of Agriculture about the locations of the wells.

Water analysis

The water samples from the study area included: soluble cations and anions, pH and ECw (EC in water).

Validation of Hydrochemistry Data

The hydrochemistry data collected in this research project was validated in two ways:

1. Used **Equation 4.3** (Al-Adamat, 2002) that reflects the ionic balance in order to validate the analysis of standard inorganic constituents.

$$B\,\% = [(\sum cations - \sum anions) / (\sum cations + \sum anions)] * 100\% \qquad 4.3$$

Applied the ionic balance analysis to samples that were collected to cations including Ca, Na, K, Mg and to anions including Cl, SO_4, HCO_3, and NO_3 where all values are summed up in meq/l^{-1}. The error should be less than 5% (Coxon and Thorn, 1989). It was found that all samples collected in March 2012 and March 2013 are valid since the residual error (B%: **Equation 4.3**) was less than 5%.

2. in order to confirm the results, the titration method was applied three times for each analysis.

4.8.2. Soil Data

in order to collect soil samples, some conditions were considered during the collection process and choice of sample locations as follows:

1. The area of farms.
2. The distances between trees and vegetables.
3. The topography of the study area.

The soil sampling was distributed to present agricultural, abandoned, and protected lands. A consistent methodology was followed in order to collect samples from different locations with the assistance of a team of the Regional Centre for Agricultural Extension and Soil testing (RCAES) in Mafraq. The purpose of this fieldwork was to collect soil samples from different locations in the study area and to chemically analyze these samples for their major anions and cations.

The samples were carefully collected in plastic bags and appropriately labeled. The soil samples were collected from three different depths:

1. Top soil representing a soil depth of 2 – 10 cm.
2. Subsoil representing a soil depth of 20 – 30 cm
3. Subsoil representing a soil depth of 40 – 50 cm.

The soil sampling was continued in February 2013 in order to collected additional samples from the previously sampled locations, as well as to collect soil samples from new locations. In addition to the previously determined soil properties, additional parameters, namely depth of the organic layer, structure, and horizon of soil and parent material were considered while collecting the soil information. In 2013, fieldwork was distributed across present agricultural, abandoned land, and protected land (**Figure 4.9, Appendix B-II**). The purpose of this fieldwork was to collect soil samples from these locations in validation of the land degradation index by assessing soil quality and structure in addition to chemical analyses of these samples.

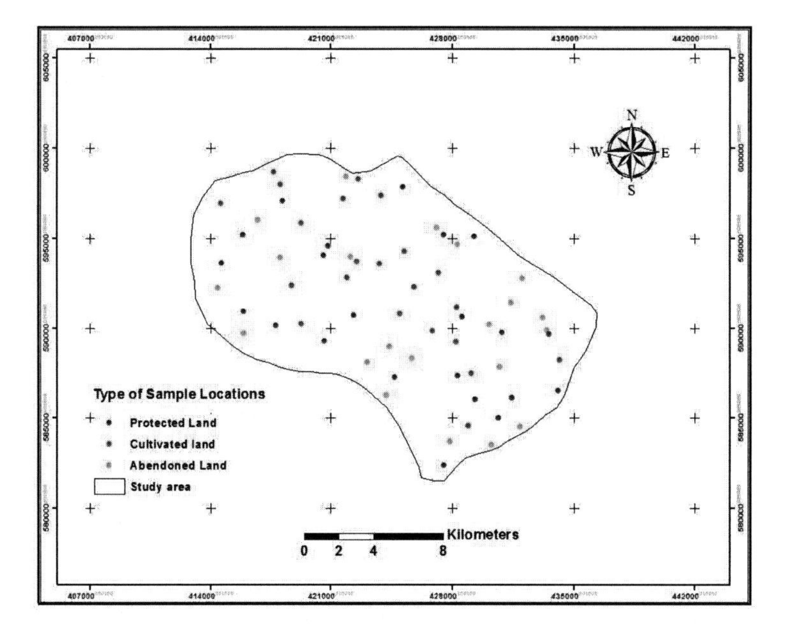

Figure 4.9: The location of soil samples for three land-use classes in 2013.

The main aims of collecting and analyzing the soil samples were to:

1. use the chemical concentration of elements in order to monitor pollution and quality of water and soil.
2. investigate the relationship between soil chemistry and groundwater hydrochemistry.
3. investigate the effect of fertilizers and pesticides on agricultural soil.
4. detection of organic materials (i.e. their percentage) in the soil.
5. investigate the relationship between soil quality and outcomes of geoinformatics.

A. Laboratory Analysis

The collected samples were air-dried and then sieved through a 2 mm sieve in order to separate the fine from the coarse soil. The analysis of the soil and water samples was carried out at the Soil and Hydrochemistry Laboratories of the Environment and Arid Regions Research Centre and at the Institute of Earth and Environmental Sciences, Jordan, respectively.

B. Soil Properties

The following substances were analyzed chemically: organic matter, soluble cations (Ca^{2+}, Mg^{2+}, Na^{+}, K^{+}) and anions (CO_3^{2-}, HCO_3^{-}, Cl^{-}, SO_4^{2-}), pH, available nitrogen (N), phosphorus (P), and electrical conductivity extract (ECe).

Table 4.6: the main analyses and methods were carried out in the laboratories of Al-al Bayt University for the soil samples and will be discussed in the procedures (work steps) in **Appendix B-III**.

Table 4.6: Main analyses and methods

Analysis	Measuring method
Na and K	Flame photometer
Organic Matter, Cl, Ca, and Mg	Titration, the organic matter was determined by the modified Walkley and Black method (Rowell, 1995) directly from the ground, dried, sieved soil extract.
Humidity	Drying
Organic Matter	Ashing
PH, EC	PH meter, EC meter + centrifuges + ratio 1:1, ratio 1:2 (water:soil)
Soil Paste	Vacuum

4.9. Descriptive Statistics

Out of the 110 soil samples, 44 from test areas in 2012 and 66 of 2013 were reserved for analysis purposes as well as being kept for validation. Descriptive statistics were applied in order to characterize the parameters of continuous environmental variables such as mean, median, and standard deviation, minimum and maximum. These samples of soil data were imported to SPSS version to describe the results statistical analysis. The univariate descriptive statistics of soil variables EC, pH, OM, soil depth, soil fertility at surface soil layer (0-30 cm) were calculated; on the other side, a *correlation coefficient* was used to measure the strength and direction of a linear relationship between two variables.

4.10. Modeling Process

The main objective is investigation of degradation resulting from impact of human activities and water in the study area, therefore in this section was described the groundwater vulnerability index and used to assessment of the risk, as well as developing a model for the investigation of land degradation; in the end, intersection process will apply between two layers to get hot-spots zones which reflect the risk degradation

4.10.1. Mapping Groundwater Vulnerability by Using Geoinformatics

Like many other countries in the world, Jordan faces serious problems related to water shortages, which negatively affect its entire development. On a global scale, Jordan is considered to be one of the four poorest countries in terms of water resources. In 2005, the Jordanian Ministry of Water and Irrigation published a crucial National Water Strategy and Policy document (MWI, 2002) which – amongst other key issues – recognized that water use already exceeded renewable freshwater resources by more than 20% and after this time Jordan started seriously to think about protect the freshwater. At the same time, Jordan has developed a very few means and search capabilities to find new water resources. The annual population growth rate of Jordan is about 2.2% (1998–2010), which is the 9th highest in the world (WAJ, 2005). Therefore, Jordan suffers from severe water shortages initiated by high population growth in addition to industrialization. Increasing demand on the existing water resources combined with declining rainfall and rising temperatures as a result of climatic changes present more serious problems.

I. Groundwater Vulnerability Indices

in order to achieve the objective of this research with respect to the identification and mapping of groundwater vulnerability, the quality of groundwater had to be predicted and evaluated by using one of the groundwater vulnerability assessments that provide flexible data within the geoinformatics environment that can be handled and interpreted. Groundwater vulnerability assessment indices have been used to collect the hydr-ogeological information in a usable form by planners, decision makers or in other studies (Jessica and Sonia, 2009). Since remote sensing cannot predict groundwater vulnerability by itself, GIS was used in this research to assess groundwater vulnerability to contamination in the study area (Ibrahim, 2010). Groundwater is intrinsically susceptible to pollution. It depends on geological and hydrogeological characteristics, but is independent of the nature of pollutants. Depending on aquifer characteristics, the most adequate vulnerability method was chosen (Navarrete *et al.,* 2008). **Table 4.7** provides literature examples of these six vulnerability methods, and the parameters of the methods used for groundwater vulnerability assessment are shown in **Table 4.8.**

Table 4.7: Groundwater vulnerability assessment methods.

Method	Aquifer	Literature
COP	Carbonate aquifers	Vı'as *et al.*, 2006
PI	Limestone and intergranular porosity aquifers	Goldscheider *et al.*, 2000
EPIK	Limestone aquifers	Doerfliger *et al.*, 1999
DRASTIC	Intergranular porosity and limestone aquifers	Aller *et al.*, 1987
SINTACS	Intergranular porosity and limestone aquifers	Civita, 1994
GOD	Intergranular porosity and limestone aquifers	Foster, 1987

Table 4.8: Methods and parameters used for groundwater vulnerability assessment.

Parameter	Methods					
	GOD	DRASTIC	SINTACS	EPIK	PI	COP
Topographic slope		X	X	X	X	X
Stream network			X		X	X
Soil		X	X	X	X	X
Net recharge		X	X	X	X	X
Unsaturated zone	X	X	X	X	X	X
Depth to water	X	X	X		X	
Hydrogeological features	X	X	X	X	X	
Hydraulic aquifer conductivity		X	X			
Aquifer thickness			X			
Land use			X	X	X	X

Groundwater vulnerability studies are expected to draw the attention of land-use planners and decision-makers to the effects of human surface activities on groundwater, especially for the identification of harmful contaminants in aquifers (Murray *et al.*, 1999). This should provide them an information basis to take action and to improve the situation. DRASTIC is an overlay and indexing method widely used to assess groundwater vulnerability to a wide range of potential contaminants. Merchant (1994) stated that DRASTIC has been used throughout the world with exceptional frequency. One of the major advantages of this model is the implementation of assessment by using a large number of input-data layers (Evans and Myers, 1990), which is believed to limit the impacts of errors or uncertainties of the individual parameters on the final outcomes (Rosen, 1994). Jordan's primary sources of water are aquifers and basin feds, which are recharged through annual rainfall. In addition, the environmental conditions in the region are marked by lack of rainfall, drought, salinity, etc. From the described situation, an urgent need can be deduced to search for techniques to study and evaluate the vulnerability of groundwater. This analysis could work for large areas; however, today the database from these techniques is not sufficient to assess the situation mainly to support the DRASTIC model.

Different environmental parameters interfere when anticipating the amount and location of impurities that may affect the aquifers, especially if the water used is extracted from aquifers for irrigation and other uses which impact on the environment. Groundwater settings, hydrological and hydrogeological conditions, land-use parameters, environmental issues, soil parameters, and other elements which may vary from one aquifer and from one area to another are used to determine the vulnerability of groundwater (Vrba *et al.*, 1994).

Several studies have used the DRASTIC model within a GIS environment, although few attempts have been made to apply the DRASTIC methodology in arid and semi-arid environments (Al-Adamat, 2002). It can also assess groundwater vulnerability through multiple considerations.

Groundwater Vulnerability Assessment with DRASTIC

In order to evaluate groundwater vulnerability for the study area by applying the DRASTIC model, multiple considerations had to be made: the DRASTIC index is considered to be one of the comparative models, and it is also flexible in its application where all the available datasets are almost reliable. The characteristics of the model are boundless, depending on aquifer properties as well as the properties of the topography, the hydrological and lithological environments. In addition, the DRASTIC method has seven parameters (**Table 4.7**). Each parameter has ratings and weights. In this research project, all DRASTIC parameters will be used without the hydraulic conductivity parameter because it lacks suitable data.

II. Ratings and Weights in DRASTIC Parameters

The DRASTIC model was chosen to investigate groundwater vulnerability in the study area. ArcView GIS was used to produce the DRASTIC index (**Equation 4.4**) - where (r) is ratings and (w) is weights and their values are deferent according to the type of parameter, with the exception of the recharge ratings that were based on **Equation 5.2** (Piscopo, 2001) instead of using the total recharge. The ratings and weights of the DRASTIC parameters for the study area were estimated from Aller *et al.* (1987) and Knox *et al.* (1993) (**Table 5.15**).

III. Equation of DRASTIC Parameters

Based on the particular conditions in semi-arid areas (such as aridity, strong fluctuations in precipitation amounts and distribution, physical properties of the area, etc.), a new vulnerability index needs to be developed that is mainly based on the gradient of the depth to groundwater, the ease of pollutants to migrate to the groundwater, soil type, net recharge equation, and land use. The rating of these conditions or variables depends on their importance for water pollution. The DRASTIC index given in the **Equation 4.4** where is considered to be an indicator for pollution potential (Merchant, 1994). The general equation of the DRASTIC index was defined by(Aller *et al.*(1987). The modified DRASTIC index - as proposed by Piscopo (2001) for the recharge ratings and weightings, by Secunda *et al.* (1998) for the land-use ratings and weightings, and by Knox *et al.* (1993) for the remaining parameter weights and ratings - is supposed to be suitable for different aquifer rock types, especially for arid and semi-arid areas with thin soil covers and a land use index (Al Adamat, 2002). The general equation for the DRASTIC index is:

DRASTIC index = DrDw + RrRw + ArAw + SrSw + TrTw + IrIw + CrCw **4.4**

Where:

Dr: Rating of the depth to the water table   Dw: Weight of the depth to the water table

Rr: Rating of the net aquifer recharge   Rw: Weight of the net aquifer recharge

Ar: Rating of the aquifer media   Aw: Weight of the aquifer media

Sr: Rating of the soil media   Sw: Weight of the soil media

Tr: Rating of the topography   Tw: Weight of the topography

Ir: Rating of the impact of the vadose zone   Iw: Weight of the impact of the vadose zone

Cr: Rating of the hydraulic conductivity   Cw: weight of the hydraulic conductivity

IV. Land-Degradation Management with Geoinformatics

Many countries in the world suffer serious problems related to desertification and degradation of lands, which negatively affects their entire development. Worldwide, Jordan is considered to be one of countries located in the semi-arid and arid areas as well as one of the four poorest countries in terms of water resources.

Furthermore, degradation will continue if human activities are not carefully controlled and managed, the factors being livestock and grazing practices, inappropriate agricultural and irrigation techniques, the marginalization of lands. the Ministry of Agriculture has started to think about developing agriculture. At the same time, they have not developed any means or search capabilities to find degradation and hot-spot areas (high degradation *and* high groundwater vulnerability). Therefore, Jordan suffers from land degradation due to using unsuitable cultivation practices and persisting periodic droughts, all of which deteriorate ecological conditions. Such a fragile ecosystem has also been manifested by non-sustainable land-use patterns and poor vegetative cover of the range land and the remaining forest batches.

4.10.2. Land-degradation Model

There are many indices to estimate desertification or land degradation, which are changes in the environmental conditions due to wind and water factors. Until now, several techniques were developed to map degradation, and degradation modeling attempts to characterize the evolution of degradation signals. So the main objective of this project was to develop an ESAs (Environmentally Sensitive Areas) model which is considered to be one of the most important environment monitoring techniques for desertification. LDD (Lands Degradation Degree) model was developed through ESAs model with a new formula by using new parameters. LDD model could be useful to identify places that suffer from land degradation through studying the merits of the deterioration in the study area and using modern geoinformatics techniques as well as lab analyses for the development of the model. One of the major advantages of the LDD model is the implementation of assessment by using a large number of input-data layers, which is believed to limit the impacts of errors or uncertainties of individual parameters on the final output.

The parameters that will use the LDD model should be quite simple, flexible, strong, and widely applicable so that they can be used in different environments. It can also add parameters in each indicator, depending on the number of variables gathered in the study area – provided it has been applied previously in order to test the accuracy of the resulting maps.

Several studies have used the degradation-indices model within a geoinformatics environment (Kosmas *et al.,* 1999; Contador *et al.,* 2009; Solaimani *et al.,* 2009). Although few attempts have been made to apply the degradation-indices methodology in arid and semi-arid environments, the LDD model is considered to be among the ones that can also assess the degradation of lands through multiple considerations.

I. Definition of Lands-Degradation Model (LDD)

The LDD model reflects the degree of degradation in specific regions where it can analyze parameters such as soil, slope, geology, organic matter, vegetation, climate, and human actions. Each of these parameters is grouped into various uniform classes according to impact of degradation by weight, and rank factors are assigned in each class.

II. Data Collection

The data-collection process is related to physical environment and land-management characteristics such as soil, vegetation, and climate data, as well as land-management characteristics (**Figure 4.10**).

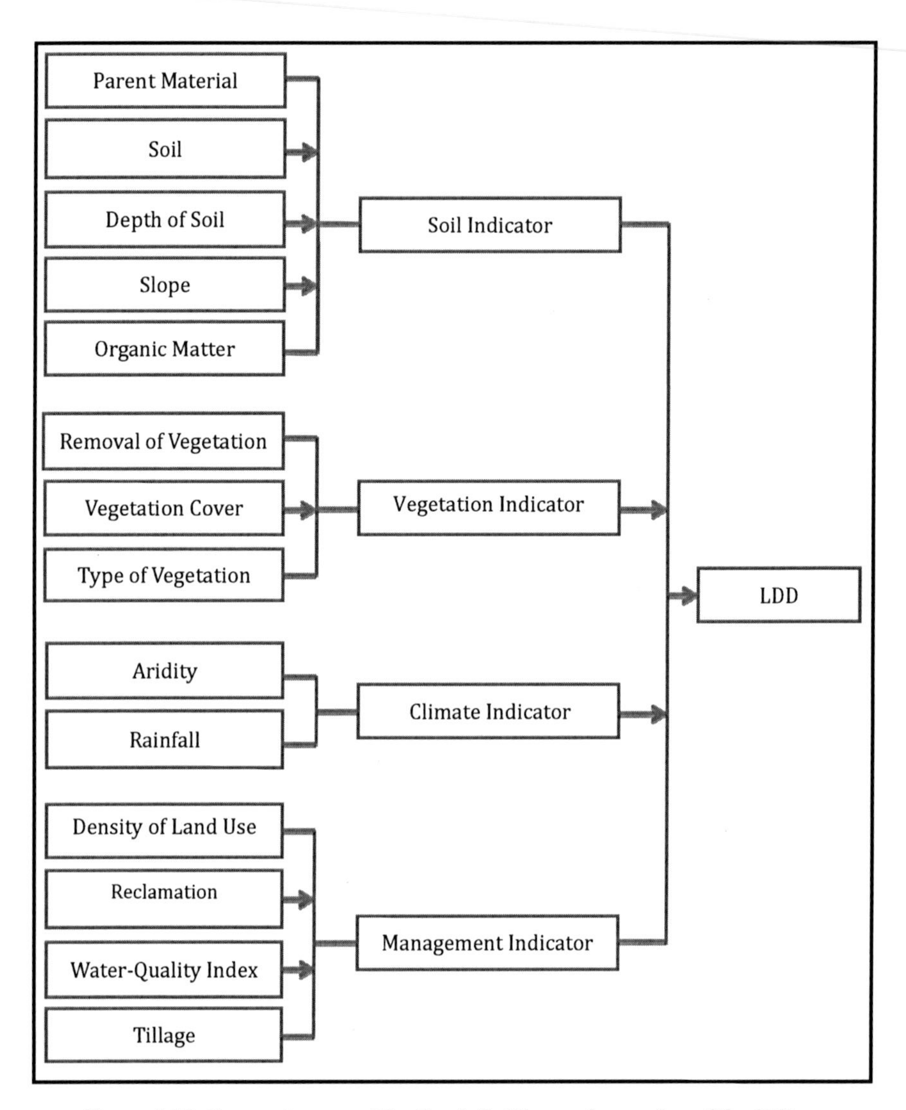

Figure 4.10: Parameters used for the definition and mapping of the LDD.

Based on the special conditions in the region – like aridity in semi-arid areas, strong fluctuations in precipitation amounts and distribution, physical properties, etc. –, a new model index needs to be developed mainly based on soil quality, vegetation, climate, as well as human actions through land management. The following explanation is given for each indicator with parameters:

A. Soil Indicator:

Soil is considered to be a controlling factor of ground ecosystems in semi-arid and dry sub-humid areas, especially when observing its impact on plant growth (DESIRE, 2007). Thus, the soil properties are considered to be the main factor in the quality of yield, nutrient content, soil biomass, organic matter, and water-holding capacity (Scherr, 1999). The characteristics change over time to "degradation" or "improvement", such as vegetation degradation when the soil has no appropriate capacity to provide water and nutrients to the largest possible area of plants. One of the most important processes of land degradation and desertification is soil erosion, which affects upland areas in particular. Land degradation and desertification are a serious threat to soil quality, productivity, and quality of future food yield. The most important parameters are:

Parent material: Parent material is considered to be a soil-forming factor affecting soil properties, plant growth, soil erosion, and ecosystem resilience (DESIRE, 2007). The soil residuals have particular physical and chemical characteristics closely linked to the parent material.

Soils derived from different types of parent materials exhibit various degrees of sensitivity to land degradation and desertification (Kosmas, 1999). Zones with soils formed on shale are normally characterized as providing large amounts of water previously stored to the plants and protecting areas from desertification because they provide high productivity. In other words, soils formed on marl are not able to protect the vegetation cover from desertification and land degradation despite their considerable depth and high productivity in normal and wet years because of their high sensitivity to water erosion (Smiel, 2002; Sokouti *et al.*, 2009).

Table 4.9 designates three classes of parent materials classified according to their sensitivity to desertification (Kosmas, 1999; Contador *et al.*, 2009).

Table 4.9: Parent-material types.

Parameter	Class	Description	Ranking
Parent material	1	Shale, schist, basic, ultrabasic, acid, conglomerate	1
	2	Granite, marble, sandstone, alluvium, siltstone, mudstone	3
	3	Clay, clayey, marl	5

Soil texture: The texture (particle-size sorting) of a soil is specified by its relative amounts of sand, silt, and clay (Tripepi, 2011). Sand, silt, and clay are the soil fractions of 2.0 to 0.05 mm, 0.05 to 0.002 mm, and less than 0.002 mm, respectively (USDA, 1987; FAO, 2006). Soil texture significantly impacts the exchange of soil, its temperature, erosion, and water-holding capacity, as well as fertility and plant productivity. Soil texture affects soil resistance to erosion.

A loam does not have natural percentages of the three classes of the soil (sand, silt, and clay), but has the characteristics of each of them: clay for water- and nutrient-holding capacities, while sand and silt provide pore space for air and water movement (Tripepi, 2011). Thus, loam soils have the best physical and chemical mix characteristics for agriculture and plant growth. **Table 4.10** shows 4 classes that were designated: sand (S), loamy sand (LS), sandy loam (SL), loam (L), silt loam (SiL), silt (Si), clay loam (CL), sandy clay loam (SCL), silty clay loam (SiCL), clay (C), silty clay (SiC), and sandy clay (SC) and classed were modified by (USDA, 1987; FAO, 2006).

Table 4.10: Types of soil texture.

Parameter	Class	Description	Ranking
Soil	1	Loam, sand clay loam, sandy loam, clay loam	1
	2	Sandy clay, silt loam, silty clay loam	2.33
	3	Silt, clay, silty clay	3.66
	4	Sand	5

Depth of soil (cm): Soil depth is important input parameter with an impact on desertification and land degradation. Spatial formats in soil depth result from complex interactions of several factors (topography, parent material, and climate, biological, chemical, and physical processes) (Summerfield, 1997; Tesfa *et al.*, 2009). The soil depth determines how far the roots can grow, as well as the available water-storage capacity (Grace, 1993). Thus, soil depth is the main factor controlling the percentage of growing plants on the surface under all climatic conditions, especially under arid or semi-arid ones.

Soil depth also exerts major control of biological productivity (Gessler *et al.*, 1995) and decrease due to soil erosion is a serious threat to soil quality as well as the quality and thickness of topsoil and the nature of the soil subsurface. Soil depth in the study field sites has been measured during site visits and in the previous study. **Table 4.11** shows four classes that have been distinguished: (a) shallow, soil depth ≤ 14 cm; (b) slightly deep, soil depth 15–30 cm; (c) moderately deep, soil depth 31–75 cm; (d) deep, soil depth ≥ 76 cm.

Table 4.11: Depths of soil.

Parameter	Class	Description	Ranking
Depth of soil (cm)	1	≥ 76	1
	2	75–31	2.33
	3	30–15	3.66
	4	≤ 14	5

Slope (%): There is a strong relationship between topography and vegetation and their effects on soil formation (SWP, 2004), due to the effect on the amount of surface runoff and the loss of soil sediment (DESIRE, 2007). And soil-moisture content is influenced by topography and the landscape aspect (Daniels *et al.*, 1987; Tsui *et al.*, 2004). Therefore, the slope percentage is obviously considered to be one of the most important factors of land degradation. **Table 4.12** shows four classes that have been distinguished.

Table 4.12: Percentage of the slope.

Parameter	Class	Description	Ranking
Slope (%)	1	≤ 6	1
	2	7–18	2.33
	3	19–35	3.66
	4	≥ 34	5

Organic matter (%): Soil organic matter (SOM) is carbon-rich material that includes plant, animal, and microbial residue in various stages of decomposition (USDA, 2001). SOM is the most important indicator of soil quality and productivity and consists of a complex and varied mixture of organic substances (Jankauskas *et al.*, 2007). SOM enhances soil function and environmental quality because it binds soil particles together into stable aggregates, thus improving porosity, infiltration, and root penetration, and reducing runoff and erosion so that the soil in arid, semi-arid, and hot, humid regions commonly contains less organic matter than the one in other environments (USDA, 2001). It also provides more water availability for plants and a lesser potential erosive runoff and agro-chemical contamination (Lal *et al.*, 1998). **Table 4.13** shows five classes of SOM that have been used for the purpose of this study after a modified soil-organic-matter test (Boyd at al., 2002): (a) < 1% mineral soil, very low organic-matter content; (b) 1–3% mineral soil, moderate organic-matter content; (c) 3.1–15% mineral soil, high organic-matter content; (d) 15–30% organic soil; (e) 30% > organic carbon soil.

Table 4.13: Percentages of organic matter in the soil.

Parameter	Class	Description	Ranking
Organic matter (%) (from weight)	1	> 30	1
	2	15–30	2
	3	3.1–15	3
	4	1–3	4
	5	< 1	5

The soil indicator (SI) is then calculated as the product of the above attributes, namely parent material, soil texture, soil depth, slope, and organic-matter conditions according to the following algorithm in **Equation 4.5**. The soil quality is then defined by using the above tables of each parameter.

Equation 4.5:

SI = [Soil texture (r*w) + Parent material (r*w) + Depth (r*w) + Slope (r*w) + Organic Matter (r*w)]

B. Vegetation Indicator:

Vegetation is a key factor that affects soils with all its activities, including degradation of land and erosion control. The vegetation protects the soil by holding it in place, promoting the availability of soil minerals, increasing water infiltration (Barry, 1996), and changing the sediment-runoff yield, which greatly helps the positive effect vegetation has on the erosion process (Mingguo *et al.,* 2007). The spatial scales difference causes a variation in the effects of vegetation on erosion and sedimentation processes (Xian-Li *et al.,* 2006) and hence is of the greatest benefit for soil conservation. The most important parameters are:

Type of Vegetation: The vegetation plays a significant role in the organization of changes in soil properties. Because the vegetation-cover type defines the specific characteristics of land uses such as olives, cereals, citrus, conifers, etc. (Kosmas, 1999), the soil surface's physical conditions are different under various vegetation-cover conditions (Vásquez *et al.,* 2010). In other words, the most important compatibility between land use and soil for vegetation types is the impact on the quality of the soil, because soil nutrients vary and change from one land-use type to another (Senjobi and Ogunkunle, 2011). This leads to soil erosion and land degradation. A similar case is the Mediterranean region suffering from land use, which leads to the removal of plants types (Dunjó *et al.,* 2004) and ultimately to land degradation. Olives have special properties and are able to adapt to droughts in the long term under semi-natural and various other conditions; they support a remarkable diversity of flora and fauna – even to a larger extent than some natural ecosystems –, providing protection for mountain areas from erosion and degradation. **Table 4.14** shows five classes of vegetation land that have been investigated for the purpose of this study (Contador *et al.,* 2009).

Table 4.14: Types of vegetation land.

Parameter	Class	Description	Ranking
Vegetation Land	1	Cultivation in Mediterranean areas, evergreen forests	1
	2	Citrus, conifers, deciduous and olive trees	2
	3	Long-living trees	3
	4	Pastoral perennial areas	4
	5	Annual agricultural crops and green land, barren land	5

Vegetation cover (%): The percentage of soil that is covered by green vegetation is defined as 'plant cover'. It can also improve soil properties and structure under the vegetation, thus the latter has direct influences on soil loss (Xian-Li *et al.,* 2006). The percentage of vegetation is estimated from vegetation indices such as NDVI, SAVI, MSAVI, and TSAVI, by using the spectral reflectance of various vegetation covers from satellite data. The vegetation cover has been measured by digitization and visiting sites, with satellite images also being used for extensive areas. **Table 4.15** features five classes of vegetation cover that have been determined for the purpose of this study: (a) < 10%, (b) 10–25%, (c) 25–50%, (d) 51–75%, and (e) > 75% (Kosmas, 1999; Contador *et al.,* 2009).

Table 4.15: Percentage of vegetation cover.

Parameter	Class	Description (%)	Ranking
Vegetation cover (%)	1	> 75	1
	2	51–75	2
	3	25–50	3
	4	10–25	4
	5	≤ 10	5

Removal of vegetation: Removal of plant cover jeopardizes the carbon cycle and decreases the percentage of organic matter in the soil, as well as resulting in a degradation of physical soil properties and structure (Albaladejo *et al.,* 1998). The removal of vegetation processes increase erosion and land degradation, particularly in the semi-arid and dry sub-humid areas that have a high sensitivity to desertification and degradation. This deteriorates the ecosystem (Lindenmayer and Burgman, 2005). Removal of vegetation is estimated relatively by measuring the ratio of annual losses of vegetation in the total surface area. This indicator is measured by using a satellite dataset of earth observations. **Table 4.16** shows three classes of vegetation cover that have been used for the purpose of this study (DESIRE, 2007).

Table 4.16: Percentages to removal of vegetation

Parameter	class	Description	Ranking
	1	Low, < 1.5	1
Removal of vegetation (%)	2	Moderate, 1.5–2.5	3
	3	High, > 2.5	5

The vegetation indicator (VI) is assessed as the product of the above vegetation characteristics in relation to land degradation by using the algorithm in **Equation 4.6.** Then the vegetation quality index is classified into three classes, defining the quality of vegetation with respect to land degradation. The vegetation quality is then defined by using the above tables of each parameter.

Equation 4.6:

VI = [Vegetation land (r*w) + Vegetation Cover (r*w) + Removal of vegetation (r*w)]

C. Climate Indicator:

Climate is an important factor impacting on vegetation growth, providing water, and causing soil erosion and land degradation. The climate change from increasing temperatures (global warming) changes water resources and growing seasons, which already begins to influence many natural systems (DPP, 2007). The dryland climate has a major impact on soil and vegetation (Sivakumar, 2007). For this indicator, the identified parameters relate to climatic conditions that can create large water deficits and affect land degradation. The most important ones are (a) the aridity index and (b) annual rainfall.

Aridity index: This aridity index measures the degree of dryness of a climate at a given location and thus defines climatic zones (Maliva *et al.,* 2012). For this study, the index by the Intergovernmental Panel on Climate Change (IPCC, 2007) was used. The arid regions are commonly perceived as receiving less than 250 mm of precipitation per year; for semi-arid areas, which are more vulnerable to desertification, the functional relationship between precipitation, potential evapotranspiration (PET), and temperature – according to a classification of climatic zones proposed by UNEP in 1997 (Robert *et al.,* 2008) – is measured by quantifying the precipitation deficit over atmospheric water demand with the aridity index (AI). This UNEP aridity index (AI) is based on:

Aridity Index = MAP/MAE 4.7

Where MAP is the mean annual precipitation and MAE is the mean annual evapotranspiration. **Table 4.17** shows the five classes of vegetation cover that have been used for the purpose of this study: (a) aridity < 0.5 (arid), (b) aridity = 0.5–1 (semi-arid), (c) aridity = 1–1.5 (sub-humid), (d) aridity = 1.5–2 (humid), and (e) aridity > 2 (per-humid).

Table 4.17: Classes for percentages of aridity.

Parameter	Class	Description	Ranking
Aridity indicator	1	> 2	1
	2	1.5–2	2
	3	1–1.5	3
	4	0.5–1	4
	5	< 0.5	5

Rainfall rate: rainfall is the most important climatic factor. In arid zones, it is usually unpredictable and variable and, of course, is volatile and seasonal. It is considered to be important for the natural ecosystem (Altwegg *et al.*, 2009); thus rainfall patterns explain the many interpretations and changes in the environment, especially in the so-called desertification and degradation processes. The arid and semi-arid areas are defined as falling within rainfall zones of less than 250 mm and 250–500 mm respectively (FAO, 1987). These areas are not suitable for cultivation.

Table 4.18 shows five classes of annual rainfall that have been distinguished based on experimental data on soil erosion obtained from previous research and other pertinent research projects on desertification: annual rainfall at (a) < 280 mm, (b) 280–650 mm, (c) 650–1000 mm, and (d) > 1000 mm.

Table 4.18: Amount of rainfall per year (mm).

Parameter	Class	Description	Ranking
Rainfall rate (mm)	1	> 1000	1
	2	750–1000	2
	3	500–750	3
	4	250–500	4
	5	< 250	5

The two above attributes are combined in order to assess the climate indicator (CI) by using the following algorithm in **Equation 4.8**. The climate quality is then defined by using the above ranges of each parameter.

Equation 4.8:

VI = [Aridity (r*w) + Rainfall rate (r*w)] **4.8**

D. Management Indicator:

Land-management practices impact on crop yields differently in the various environmental zones (Samuel, 2010); furthermore, the types of land-management practices used differ across different ecological zones. In many areas, unregulated activities seriously affect biodiversity and land cover (vegetation), threaten public health and water resources (Darwish, 2009), reduce biological productivity and usefulness of land resources, and what stems from human activity in environmental management is one of the many factors responsible for land degradation (Gretton and Salma, 1997).

Unsustainable land-management practices are exacerbated by the weakness or absence of adequate policies and regulations (Darwish, 2009). Therefore, the impacts of land degradation can be overtaken by taking into account and applying appropriate measures and land-management practices. Based on that, the management of agricultural lands should be improved by enhancing the practices in all zones and an amendment of land-management practices that are not used but have potential to improve agricultural productivity and soil, and by following up on integrated approaches for sustainable land management. In this study, the following land-management indicators for combating desertification and land degradation have been considered: (a) density of land use, (b) reclamation of affected areas, (c) water-quality index, and (d) tillage.

Density of land use: Land-use intensity is indicated by types of land uses, rangelands, cultivation, forests, and other uses. Density of land use in agricultural lands are related with degree of cultivation by mechanized applications, use of fertilizers and pesticides as well as irrigation system (DESIRE, 2007; Truscott, 2012), tracking for several applications that will improve and increase food production without further depleting soil and water resources (World Bank, 2006; FAO, 2011), and improving productivity and soil quality. The land-use intensity for cropland has been assessed by characterizing the frequency of irrigation, the degree of mechanization, the use of agrochemicals and fertilizers, the crop varieties used, etc.

Table 4.19 shows three levels of land-use intensity that can be distinguished for agricultural areas (DESIRE, 2007; Kosmas, 1999):

• *Low* land-use intensity (extensive agriculture): fertilizers and pesticides are not applied and mechanization is limited. In the case of seasonal crops, one crop is cultivated per year or the land remains fallow.

• *Medium* land-use intensity: improved varieties are used, insufficient fertilizers are applied and mechanization is restricted to the most important tasks such as sowing.

• *High* land-use intensity (intensive agriculture): improved varieties are used. Application of fertilizers and cultivation is highly mechanized.

Table 4.19: Levels of land-use intensity.

Parameter	Class	Description	Ranking
Density of land use	1	Low intensity	1
	2	Medium intensity	3
	3	High intensity	5

Reclamation of affected areas: Land reclamation refers to the active of uses which could reduce the problems of land degradation (Jimoh *et al.,* 2011), soil erosion, acidification, heavy metal contamination, and – most importantly – salinization, especially in areas sensitive to desertification and land degradation. Since land degradation is a global problem, soil salinization has attracted global attention as a main form of it (Zhao and Meng, 2010), which is reflected in the land-reclamation degree in regions with an applied degradation index (Zhao and Meng, 2010); it is modified in **Equation 4.9**, which depends on DN values of salt wasteland that has been assessed qualitatively by using remote-sensing data and analysis samples in the fieldwork. **Table 4.20** shows three levels of reclamation of affected areas.

Degradation index = 255 - (B2+B3)/255 + (B2+B3) **4.9**

Table 4.20: Levels of reclamation of affected areas.

Parameter	Class	Description	Ranking
Reclamation of affected areas	1	Reclamation (Low salinity) – DI < 30	1
	2	Moderate reclamation (Moderate salinity) – DI = 30–60	3
	3	No reclamation (High salinity) – DI > 60	5

Water-quality index: The water-quality index is considered to be a most important standard in assessing the quality of irrigation water in order to avoid or at least reduce impacts on agriculture. It has been used for environmental decisions and the application of irrigation water protection (Muthanna, 2011); thus, the purity of water is also an important factor in agriculture (Jezierska *et al.,* 2011), and the quality of irrigation water directly affects the quality of the soil (Muthanna, 2011). These factors affect soil nutrients and change their properties, thus leading to soil degradation and erosion.

The quality of water for irrigation reflects inputs from soil, pollutant sources and atmosphere; it also depends on the nature, composition of the soil, sub-soil, climate, topography, etc. (Dhembare, 2012). Thus, irrigation water depends on dissolved salts such as Na, Ca, Mg, and HCO_3 in water; the ratio of their concentration affects the quality of water for irrigation (USEPA, 1974), increases the concentration of salts in it, and changes the soil quality (Dhembare, 2012). The sodium-absorption-ratio (SAR) factor has been used to evaluate water suitability for irrigation. This is considered to be one of the most common factors that influence the normal rate of water infiltration (Muthanna, 2011).

The SAR value of irrigation water quantifies the relative proportions of sodium (Na^{+1}) to calcium (Ca^{+2}) and magnesium (Mg^{+2}) and is computed as **Equation 4.10**:

$$SAR = [(Na^+)/\sqrt{(((Ma^+) + (Ca^+))/2)}] \qquad 4.10$$

The classes in **Table 4.21** have been defined as follows (FAO, 2006): (a) SAR < 3 – none, (b) SAR = 3-9 – slight to moderate, and (c) SAR > 9 – severe.

Table 4.21: Values of the SAR in irrigation water.

Parameter	Class	Description	Ranking
Water-Quality Index	1	SAR < 3	1
	2	SAR = 3–9	2
	3	SAR > 9	5

Tillage (cm): Tillage is a mechanistic modification of soil construction through processes that modify soil construction such as rebound, cutting, beating, etc., which depends on the properties of the tillage process (depth and width of disturbance etc.) and the properties of the soil (structure, texture, friability, etc.) (NRCS, 2011).

The mechanical tillage of soil affects the productivity elements that derive from the different plants and factors of vegetation, mostly on water and nutrients of soil (Rusu *et al.,* 2011), and affect the physical, chemical, and biological properties of soil; the degradation also influences the quality of soil through the disintegration of biological nutrients (Gilley and Doran, 1997). On the other hand, the decline of tillage practices helps prevent structural soil degradation and erosion losses (Ferrara *et al.,* 2012); especially deep tillage improves the soil-moisture content (TNAU, 2008). Thus, deep tillage distinguishes semi-arid and arid areas from humid areas by their amount of rainfall, with deep tillage being more convenient in semi-arid and arid areas.

Table 4.22 shows classes that have been defined as (DESIRE, 2007): (a) shallow, tillage depth < 20 cm; (b) moderately deep, tillage depth 20–40 cm; (c) deep, tillage depth > 40 cm, and (d) no tillage.

Table 4.22: Tillage depths in soil.

Parameter	class	Description	Ranking
Tillage Depth (cm)	1	> 40	1
	2	20 - 40	2.33
	3	< 20	3.66
	4	No tillage	5

The management index (MI) is assessed as the product of the above parameters by using the following algorithm (**Equation 4.11**). Then the management quality is defined by using the above tables of each parameter.

Equation 4.11:

VI = [Density of land use (r*w) + Reclamation of affected land (r*w) + Water quality (r*w) + Tillage (r*w)]

6.5. Weights and Ranking in the LDD Model Parameters

The model in this work has four indicators; each contains a number of parameters. They were assessed by computing a vulnerability model resulting from values of deferent variables arranged as follows: soil, vegetation, climate, and management; the values weight and ranking were assigned to them, based on the importance for and relationship to land degradation, from 1 for the lowest to 5 for the highest one.

More parameters can easily be extracted from fieldwork and bioinformatics databases at various scales, the developed land degradation model was calculated by equation 6.7:

$$\text{LDD Model} = \sum P\, w*r \qquad 4.12$$

Where P are parameters of indicators, R is ranking and W is the mean weight. In order to make the spatial calculation, a raster calculator should be represented at the same pixel size. In the end and in the final stage of developing, land degradation management was further subdivided into classes from very low to very high. Data and maps were calculated, integration, management, and processing were performed by means of multi-geoinformatics software. All maps of the above parameters will be saved in raster format. In order to retrieve the final vulnerability map, the images were processed in

ArcView by multiplying the raster values of each output map (degradation indicators) with their corresponding vulnerability-weighting coefficients. Thus, they are easy to handle in the clarification and mapping of land degradation.

The ranking values have been set in accordance with the highest number of classes in the indicator so that the low values represented regions indicating that there is no land deterioration or low degradation, while high values represented areas with high degradation, indicating there is evidence for land deterioration, and weights refer to the degree of influence of each indicator on the deterioration where two scenarios exist to determine values of weight for each indicator (**Table 4.23**). They have been developed with a questionnaire handed out to all stakeholders, i.e. farmers, researchers (geologists, agriculture engineers, and soil and vegetation specialists), and people who were interested in land degradation as well as in the results of our fieldwork. Based on these questionnaires and the fieldwork, the values of weights were developed.

Table 4.23: Scenarios for the values of weights.

	Weight			
Scenario	**Climate**	**Soil**	**Vegetation**	**Management**
Arid	3	5	3	4
Non-arid	3	4	5	2

4.11. Intersect Process

The marge process includes using the intersect tools in the GIS environment between groundwater vulnerability (modified DRASTIC) and land degradation model in order to determine the hot-spot areas that are considered to feature high degradation. Then, the models could be validated and the result given (**Figure 4.11**).

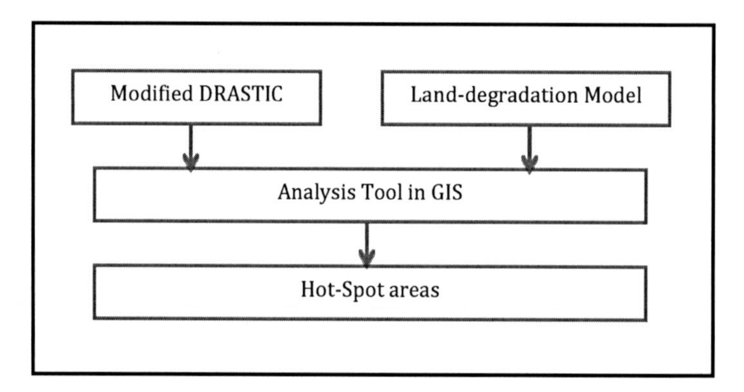

Figure 4.11: Intersect Process between groundwater vulnerability (modified DRASTIC) and land-degradation model.

Results

5.1. Introduction

In this chapter, the results obtained from data evaluation are presented: It consists of the extraction of thematic information from the satellite imagery and the detection of change in the time series data. Further analysis for the soil and water samples as well as subsequent statistical analysis were done to determine the required parameters to modify groundwater vulnerability index (DRASTIC index), in addition to develop an ESA (Environmentally Sensitive Areas) model through land degradation model and validation of results.

5.2. Image Analysis

The results showed that there is a change in land use, resulting from the fact that the study area was subjected to a decrease in the tree and vegetation coverage due to bare soil extending towards agricultural lands, which cover vast parts of the study area. **Figure 5.1** (**A, B**) shows the classification results of the object-oriented interpretation for land use and coverage in the study area by using low-resolution satellite data (TM) from 1990 and 2010 respectively. In the bottom of study area many semi-arid areas appeared where doesn't exist before, and the general features of the areas changed, in other hand agricultural areas declined in the years after 1990. While the analysis show that agriculture still semi-stable in its center of study area.

Change Detection

In this study, a twenty-year span (1990 to 2010) was investigated, which is a short time span in a long history of land coverage and use. This period was chosen based on its availability of data for the study area. In the last two decades, the Eastern part of Jordan has been amongst the most dynamic and most fundamentally changing areas. The area has been affected by many natural and human activities. Results obtained from current and previous land-coverage studies basically summarize the main factors responsible for these changes, which include the increase in the depreciation of land use and the spread of artesian wells; the land-use and land-coverage change detection was assessed by using a traditional post-classification, cross-tabulation approach. **Table 5.1** shows the percentage of change of the different land-use classes in surface areas.

Table 5.1: Percentage of change of the different land-use classes in surface areas.

Land Use	Area in km^2		Change 1999 – 2010	%
	1990	2010		
Bare soil	93	146	+ 53	+ 20.7
Trees	53	22	- 31	- 12.1
Vegetation	110	88	- 22	- 8.59
Total	256	256		

Changes in the land cover were unidirectional for all areas. The absolute rates of change for bare soil were higher during the period 1990 – 2010. A significant decrease in the areas of agricultural land and pastures could be noted.

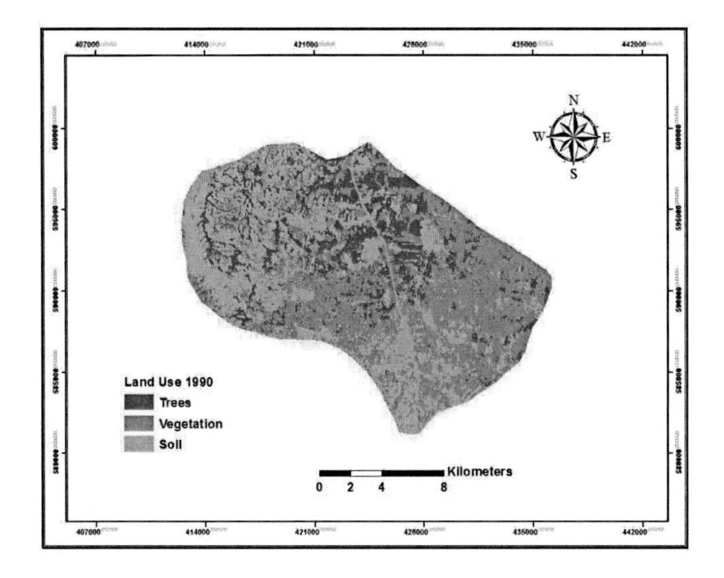

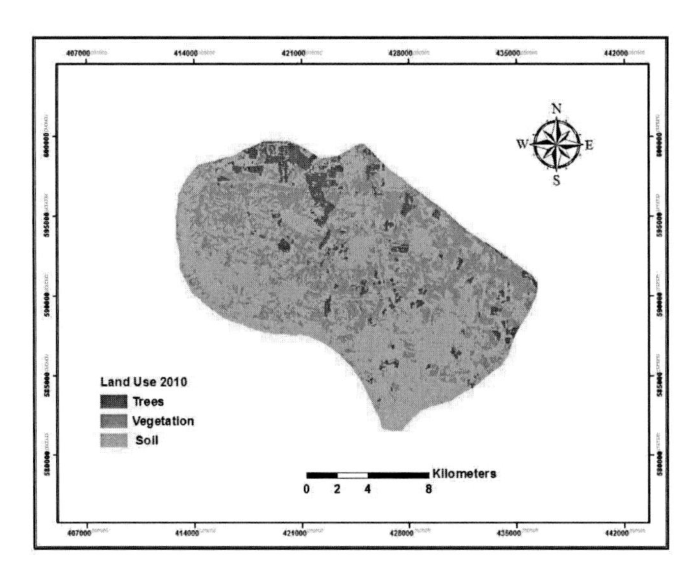

Figure 5.1 (A and B): Land-use classification maps for 1990 and 2010.

Tables 5.2, 5.3, and **5.4** represent correct classifications. The assessment of the accuracy was done for the whole study area in order to investigate the classifications and these changes, where the overall classification accuracies (sum of diagonal element divided by grand total, expressed as percentag3e) were 84.47% and 77.74% for 1990 and 2010 respectively. On the other hand, the overall kappa statistics reached 0.7992% and 0.7291% for 1990 and 2010 respectively. This means there is matching between results of classification and ground data.

Table 5.2: Accuracy assessment for land-use and land-cover TM data for study area in 1990.

Class	Reference total	Classified Number total	Number correct	Producer accuracy	User accuracy
Soil	38	44	33	87	75%
Trees	7	9	7	100	78%
Vegetation	55	50	47	81	94%

Table 5.3: Accuracy assessment for land-use and land-cover TM data for study area in 2010.

Class	Reference total	Classified Number total	Number correct	Producer accuracy	User accuracy
Soil	35	46	35	100	76%
Trees	7	10	6	85.71	60%
Vegetation	50	55	45	90	82%

The user accuracy was determined by dividing the number of correctly classified objects by its row total. It indicates the probability that an object from the land-use and -cover map actually matches the one from the ground-truth data. On the other side, the overall kappa statistics reached 0.7992% and 0.7291% for 1990 and 2010 respectively.

5.3. Field Survey and Laboratory Analysis

Understanding the causes of in-field variability in types of soil is crucial to achieve successful land management. Investigating the diverse and complex factors that control land quality requires a holistic approach.

Surface soil samples are needed for all land-use type. Fertilizer recommendations for all nutrients except nitrogen are based on the crop to be grown and soil tests of the surface samples. These were used for determining soil pH, electrical conductivity (ECe), organic matter (O.M.), AH Horizon, texture and structure of the soil, and soil moisture.

5.3.1. Physical and Chemical Properties of the Soil

5.3.1.1. Physical Properties

A. Soil texture

Seventy-two soil profiles were investigated and a small-scale soil map (1:25,000) was consulted in order to obtain an accurate representation of the soil texture in the Yarmouk basin and the study area was one of them (Ministry of Agriculture, 1994). The study area included two soil texture (silt clay loam and silt loam), and there were several soil types of textures. **Table 5.4** summarizes these types.

Table 5.4: Soil textures within each unit of the study area (MoA, 1994).

Soil texture	Unit			
Silt clay loam	SAB	THE	BIR	-
Silt loam	BUQ	THR	BIR	BUR

According to our present knowledge through fieldwork and soil analysis, based on all previously published soil studies, available data, and according to USDA Soil Taxonomy (Al-Qudah *et al.*, 2001), soil textures could be identified in Jordan. Since these textures within each soil unit had no definite border, it was difficult to design a map that represented them and is also provided by the Ministry of Agriculture **Table 5.4**; so the percentages of sand, clay, and silt were appropriate to the soil-unit textures. According to the USDA soil classification and with soil analysis and fieldwork, this research categorized the soils of the study area **Chapter 3, Section 3.2.6** into two classes **Figure 3.5**: silt clay loam and silt loam.

B. Soil Moisture

The results of the moisture content for the soil found differences between types of land for several samples in the study area. Where soil moisture content of cultivated land was lower in 2012 than in 2013, this might be the result of a possible increase in the use of water (excessive irrigation) on cultivated lands in 2013. **Table 5.5** shows the soil moisture contents in 2012 and 2013, where the maximum of the soil moisture had gradually increased for cultivated lands from 17.6% to 18.3%, while the maximum on the abandoned land had decreased from 11.2% to 10.7%; this might be due to the fact that both sites received no recent irrigation.

Table 5.5: Soil moisture contents (%), with mean and standard deviation by profile.

Soil moisture content	Protected		Cultivated		Abandoned	
Years	2012	2013	2012	2013	2012	2013
Min	-	9	10.2	11.5	6.4	6.3
Max	-	15.2	17.6	18.3	11.2	10.7
Mean	-	11.1	14.2	14.4	8.8	8.9
SD	-	1.5	1.86	1.8	1.3	1.6

The minimum and maximum SMC values of the respective test samples suggest that the study area doesn't generally have a homogeneous texture and the high SMC values are a result for the main colloidal particles (clay) in the soil.

C. Organic Matter

The percentage of organic matter formed 1.5 % of total size soil sample in the arid and semi-arid lands. Therefore, the results of organic matter values for the types of land in the study area in 2012 and 2013 were less than 1.5% **Appendix B-II**. In the 2013 samples, it was found that some lands included values ranging between 1.5% and 2.2%; four samples were taken from the cultivated lands (H14A, H47A, H55A, and H65A), depending on foundations in correct agricultural practices; another two samples were taken from abandoned lands (H38A and H59A), i.e. from farms which were abandoned a year ago; and five samples were taken from protected lands (H31A, H62A, H63A, H75A, and H57A) i.e. locations that had not been used before. The organic matter was less than 1% on all lands within the study area; this result is decreasing due to pressure from agricultural and human activities. Protected lands become cultivated and eventually abandoned. **Table 5.6** shows mean and standard deviation as well as the minimum and maximum of organic matter. This means that there is breakdown of organic matter on these lands, and the variability in organic matter in the study area is a result of agricultural crop residues, manure added to the soils, and farming practices.

Table 5.6: Organic matter percentage, with mean and standard deviation by profile.

Organic matter	Protected		Cultivated		Abandoned	
Years	2012	2013	2012	2013	2012	2013
Min	-	0.54	0.76	0.51	0.5	0.41
Max	-	2	1.9	2.2	1.3	1.3
Mean	-	1.07	1.21	1.19	0.95	0.94
SD	-	0.37	0.33	0.54	0.26	0.29

D. Soil structure

Two types of soil structure were recorded through fieldwork in the study area. First, moderate medium granular and strong medium sub-granular blocky existed in different proportions, such as 32% on protected, 59% on cultivated lands, and 27% on abandoned lands. While the second type – strong fine granular, hard, and sticky soil – existed in different and somewhat higher proportions than the first type, for example 68% on protected, 41% on cultivated, and 73% on abandoned lands. The existence of both types might be a result of climatic conditions and a lack of agricultural land reclamation. **Table 5.7** shows the soil horizons with types of soil structure.

E. AH horizons

The AH horizon of the soil was similar within the study area, as well as at a depth of 10 cm, where an A horizon (Ap) could be found; plowing or other anthropogenic disturbances of the surface layer as well as organic matter appeared to be less than 1.5% of the constituent materials, and the soil lacks the distinctive characteristics of E or B horizons, whereby the elements of Ap horizon are clearly influenced by human activities resulting from cultivation, pasturing, or similar disturbances. **Table 5.7** shows an overview of the coexisting soil characteristics (structure, horizon, and soil texture).

Table 5.7: Soil characteristics (structure, horizon, and soil texture).

Study area	Horizons	Ap		
	Structure	M-M granular	S-granular	
	Texture	Silt loam	Silty clay loam	Silt loam

Ap: plowing or other anthropogenic disturbance of the surface layer.
M-M granular: moderate medium granular and strong medium sub-granular blocky.
S-granular: strong fine granular, hard, and sticky.

5.3.1.2. Chemical Properties

A. Acidity-Alkalinity (pH)

Soil reaction or pH is a valuable diagnostic measurement. It indicates whether the soil is acidic (pH 1-6), neutral (pH 7), or basic (pH 8-14). The pH is defined as the logarithm of the reciprocal of the hydrogen ion concentration. The equation is:

$$pH = -\log_{10}[H+]\ (mol/l) \qquad 5.1$$

The pH value ranges from 7.40 to 8.5 in the study area and the availability of soil nutrients to plants is optimal with a pH range of 6.5 to 7.0. Where the pH exceeds this value, it consequently has a negative effect on nutrient availability.

B. Electrical conductivity

Table 5.8 shows the SD for the EC, which was 6.3 in 2013 for abandoned lands, while it was in 2012 (SD = 7.94). This indicates that there is a high variability in EC values from one site to another within the study area. The standard deviation for each soil profile also appears to be variable on the same land even if it has the same use. Furthermore, it emerges from **Table 5.8** that the maximum of EC values for soil extracts collected on cultivated fields in 2013 were higher than maximum EC values on protected land (50.2 > 18.21). This might indicate that the agricultural activities in 2013 led to an increase in soil salinity.

While EC values seem to be similar on abandoned and protected lands, this could be a result of rainfall that either washed off part of the topsoil with runoff or recharged groundwater. Over time, the high values of EC affect the neighboring middle and low ones **Figure 5.2**. This can lead to land deterioration, degradation of soil quality, and pollution of groundwater.

Table 5.8: EC values, mean, median, and standard deviation of soil-water extracts (mS/cm^{-1}) in March 2012 and May 2013.

EC	Protected		Cultivated		Abandoned	
Years	2012	2013	2012	2013	2012	2013
Min	-	0.23	0.25	0.25	0.4	0.4
Max	-	18.21	44.6	50.2	17.4	19
Mean	-	5.92	3.54	3.78	5.08	5.12
SD	-	6.46	9.24	10.3	7.94	6.3

Also, the highest values of EC are located in the southwest of the study area, which means this area suffers from high salinity that is described as unsuitable for cultivation or other use because it has exceeded the allowable limit of appropriate standards for agriculture **Chapter 2, Section 2.3, Table 2.1.**

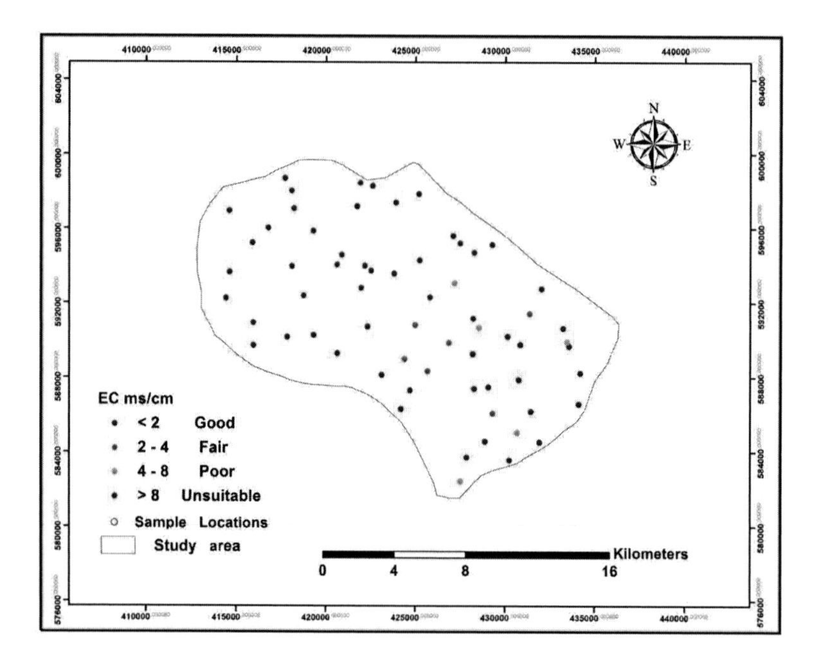

Figure 5.2: Electrical conductivity (EC) (ms/cm^{-1}) in the study area (March 2013).

Generally, saline soils reduce agricultural crop growth. The high soil content increases the osmotic potential of the soil solution and prevents crop uptake of water.

C. Sodicity (Sodium Absorption Ratio = SAR)

Table 5.8 shows the SD for the SAR as 6.15 in 2013 for abandoned lands, while it was 6.01 in 2012. This indicates that there is a high variability in SAR values from one site to another within the study area. The standard deviation for each soil profile also appears to be variable on the same land despite identical use. Furthermore, it emerges from **Table 5.9** that the maximum SAR values for the soil extracts from samples collected on cultivated fields in 2013 were higher than the maximum of SAR values on protected and abandoned lands (25.92 > 23.65 > 20.57, respectively). This might indicate that the agricultural activities had led to an increase in soil salinity by 2013, which confirms the EC values. The high values of SAR affect the neighboring middle and low values over time.

Figure 5.3 shows the impact on the growth of plants, which leads to degradation as well. This reflects the ratio of elements to enter into the sodicity equation – Na, Ca, and Mg – and the consistency with the sodium values observed **Appendix B-II**.

Table 5.9: The SAR, with mean and standard deviation by profile (March 2012 and March 2013).

SAR	Protected		Cultivated		Abandoned	
Years	2012	2013	2012	2013	2012	2013
Min	-	1.1	0.93	2.9	6.1	11.10
Max	-	19.7	24.8	19	19.6	16.20
Mean	-	7.71	9.01	11.08	12.08	14.03
SD	-	6.10	6.88	4.98	4.1	1.3

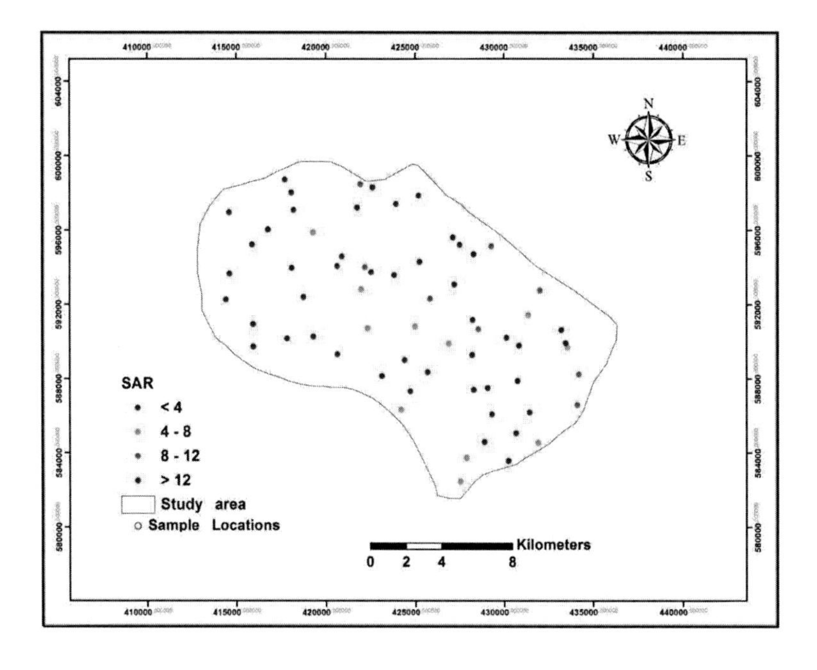

Figure 5.3: SAR in the study area (March 2013).

5.3.2. Chemical Properties of the Groundwater

In order to determine and to investigate data collected for the water quality of the limestone aquifer in March 2013, several analyses and interpreting methods had to be used, such as the spatial and vertical distribution of quality parameters, combined with the hydrogeological distribution, chemical composition, and ion ratios, scatter diagrams and graphs of time series (Stuyfzand, 1986; Da'as and Walraevens, 2010). Through the results of these data-application methods, the following interpretations emerged.

A. PH

The pH results that varied between 7.14 and 8.3 (SD = 0.181) for the samples of groundwater in private and government irrigation wells in 2012 appear in **Table 5.11**. It also shows that the pH of drinking wells varied between 7.2 and 7.89, with a standard deviation of 0.118. The minimum pH value in the irrigation wells for 2012 was higher than the minimum pH value recorded in 2013. **Figure 5.4** illustrates the pH value within the study area in March 2013. It shows that the minimum pH value was found in wells located in the central part of the area, while the highest ones were recorded in its eastern and southern parts.

Table 5.10: pH values of groundwater samples of 2012 and 2013.

pH	Irrigation wells (2012)	Drinking wells (2012)	Irrigation wells (2013)
Min	7.39	7.26	7.14
Max	7.97	7.7	8.3
Mean	7.62	7.42	7.48
Median	7.57	7.3	7.39
SD	0.181	0.118	0.301

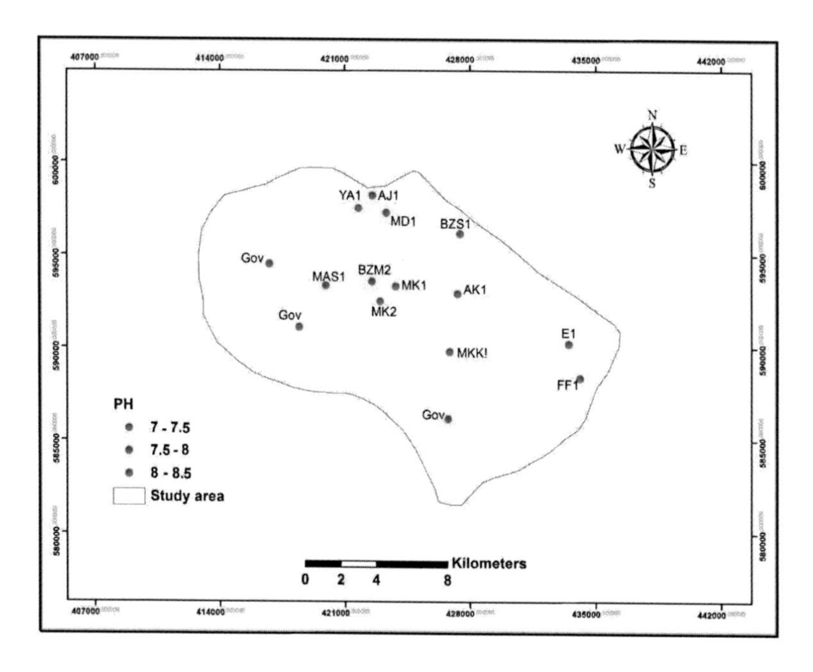

Figure 5.4: pH values within the study area (March 2013).

B. Electrical Conductivity

Results of electrical conductivity (EC) for the samples of groundwater in the private and government irrigation wells in 2012 appear in **Table 5.10**. These varied between 580 and 885 µS cm^{-1} (SD = 117.9). In 2013, there was a change in values of EC, thus the standard deviation value changed to 116.6. While the EC values in drinking-water wells varied between 295 and 506 µS cm^{-1} (SD = 63.8), the median groundwater value for EC in all wells was above 350 µS cm^{-1} both in 2012 and 2013.

Figure 5.5 shows the distribution of EC values in the wells of the study area. It shows that EC in most of the wells was between 590 and 994 µS cm^{-1} in 2013. Only one well had an EC value of more than 900 µS cm^{-1} (MD1) – in the northern part of the study area.

Table 5.11: EC values of groundwater samples of 2012 and 2013.

EC (mS cm^{-1})	Irrigation wells (2012)	Drinking wells (2012)	Irrigation wells (2013)
Min	580	295	571
Max	885	506	994
Mean	670	345	698.25
Median	640.5	385	665.5
SD	117.9	63.8	116.3

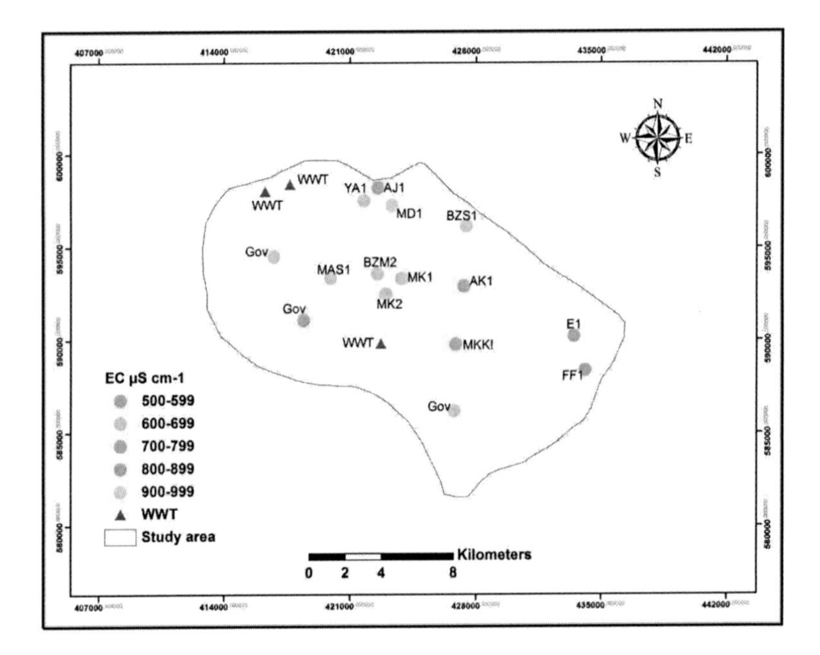

Figure 5.5: EC values (µS cm^{-1}) within the study area (March 2013).

C. Major Cations

The **Figure 5.6** show one of the major cations results for the samples of groundwater in the drinking-water and irrigation wells in 2012 and 2013. **Table 5.12** illustrates their minimum and maximum, mean, median, and standard deviation values as well as the concentrations of calcium (Ca), magnesium (Mg), phosphorus (K), and sodium (Na). The Na values varied between 101 and 111, while all the values of K did not change and were found to be equal to 1 in 2012 and 2013..

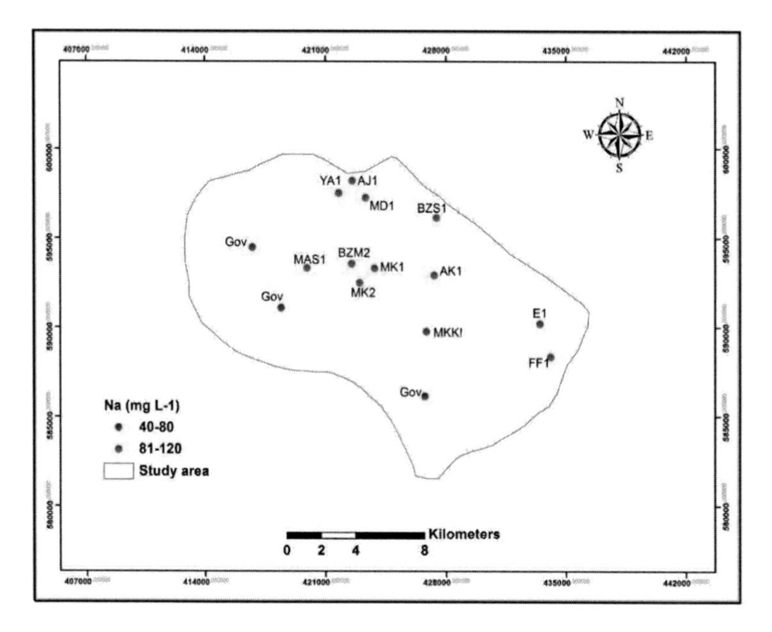

Figure 5.6: Sodium (Na) concentration (mg l^{-1}) in the study area (March 2013).

Table 5.12: The major cations in groundwater samples of 2012 and 2013.

Function	Irrigation wells (2012)				Drinking wells(2012)				Irrigation wells (2013)			
mg/l	Ca	Mg	Na	K	Ca	Mg	Na	K	Ca	Mg	Na	K
Min	52	20	96	1	12	3	40.5	1	53	18	101	1
Max	149	79	119	1	18	17.9	80.6	1	165	83	111	1
Mean	110.6	56.9	107.3	1	15	9.9	58.6	1	111	56.7	107.8	1
Median	117.5	108	108	-	15	9	56	-	114.5	61.5	107.8	-
SD	29.1	17.4	2.62	-	2.44	6.12	16.4	-	31.7	18.6	2.76	-

D. Major anions

Table 5.13 illustrates results of major anions for the samples of groundwater in the drinking-water and irrigation wells in 2012 and 2013. It also illustrates their minimum and maximum values, mean, the median, and standard deviation as well as the concentrations of bicarbonate (HCO_3), carbonate (CO_3), and sulphate (SO_4). Chloride (Cl) concentration was also found to be higher in groundwater than on the farm between 2012 and 2013. Furthermore, it was found that the maximum concentration of bicarbonate (HCO3), carbonate (CO_3), sulphate (SO_4), and chloride (Cl) in groundwater lay within the published concentrations of the same elements in natural water in 2012 and 2013.

Table 5.13: The major anions in groundwater samples of 2012 and 2013.

Function	Irrigation wells (2012)				Drinking wells (2012)				Irrigation wells (2013)			
mg l^{-1}	HCO_3	CO_3	SO_4	Cl	HCO_3	CO_3	SO_4	Cl	HCO_3	CO_3	SO_4	Cl
Min	82.2	-	20.8	45.1	80.2	-	20	22.6	79.6	-	29.6	30.2
Max	135.9	-	149.6	121.3	117	-	46.1	54.2	135.6	-	165.4	134.5
Mean	103.6	-	93.6	86.2	97.6	-	34.4	40.4	108.8	-	94.3	86.9
Median	111.3	-	98	88.4	95.6	-	37.3	44.2	109.5	-	95.1	89.3
SD	19.8	-	44.4	25.3	15	-	10.8	13.2	18.89	-	46.9	30.2

Above, the values of major anions in groundwater samples of 2012 and 2013 are shown. The cause of the disparity in values is attributed to potential sources for major anions as follows:

1. *Bicarbonate* (HCO_3) *sources*: Bicarbonate in the groundwater may have originated from water filtered through the soil.
2. *Carbonate* (CO_3) *sources*: Carbonate was not detected in any well in 2013.
3. *Sulphate* (SO_4) *sources*: The most likely source of sulphate in the study area is the oxidation of sulphide minerals. Also, SO_4 could be sourced from the dissolution of gypsum ($CaSO_4$) that could reach the groundwater aquifer with the recharge water of a Na:Cl ratio of 1:1.
4. *Chloride* (Cl) *sources*: An attributed table source of chloride in groundwater in the study area through recharged water is where water evaporates from the surface, leaving a crust of principally halite and gypsum. Heavy rain dissolves this crust and allows some infiltration of water with an Na:Cl ratio of 1:1 to recharge the aquifer (Al-Adamat, 2002; Drury, 1993).

E. Sodicity (SAR)

Table 5.14 illustrates results of SAR for the samples of groundwater in the drinking-water and irrigation wells in 2012 and 2013; it also illustrates their minimum and maximum values, the mean, median, and standard deviation as well as the concentrations of SAR, which were also found to be higher in groundwater than in farm water between 2012 and 2013.

Table 5.14: SAR concentration in groundwater samples of 2012 and 2013.

	Min	Max	Mean	Median	SD
Irrigation wells (2012)	6	18	12	11	2.03
Drinking wells (2012)	2	3	2	2	1.02
Irrigation wells (2013)	6	13	11	9	1.04

5.4. Validation of Hydrochemistry Data

The hydrochemistry data collected in this research project was validated in three ways:
1. Used **Equation 5.2** (Al-Adamat, 2002) that reflects ionic balance in order to validate the analysis of standard inorganic constituents.

$$B\ \% = [(\sum cations - \sum Anions) / (\sum cations + \sum Anions)] * 100\% \qquad 5.2$$

2. Applied the ionic balance analysis to samples that were collected to cations which include Ca, Na, K, Mg, as well as to anions that include Cl, SO_4, HCO_3, and NO_3 where all values are summed up in meq/l^{-1}. The error should be less than 5% (Coxon and Thorn, 1989). It was found that all samples collected in March 2012 and March 2013 are valid, since the residual error (B%: **Equation 5.2**) was less than 5%.

3. In order to confirm the results, the titration method is used three times for each analysis.

5.5. Salinity and Alkalinity Hazard on the Soil

The irrigation of agricultural crops depends on the groundwater, and the analysis hinted at a salinity risk directly related to the quantity of salts dissolved in the irrigation water. All irrigation water contains potentially harmful salts, and nearly all the dissolved salts are left in the soil after the applied water is lost due to evaporation from the soil or through transpiration by the plants. Unless the salts are leached from the root zone, they will sooner or later accumulate in quantities which will partially or entirely prevent the growth of most crops.

The most influential water quality guideline on crop productivity is the salinity hazard as measured in electrical conductivity (EC). The primary effect of high EC of water on crop productivity is the inability of the plant to compete with ions for water in the soil solution (physiological drought). The higher the EC, the less water is available to plants, even though the soil may appear wet. Because plants can only transpire "pure" water, usable plant water in the soil solution decreases dramatically as EC increases.

Finally, analysis data remain limited and restricted unless they are used with another method to get clearer, more realistic and comprehensive outcomes for the fieldwork data; therefore, modeling was used to present the results of data.

5.6. Groundwater and Land-Degradation Modeling

The factors should be described that include groundwater and land degradation in order to get to the final results.

5.6.1. Groundwater

In order to get to the resulting DRASTIC index, its seven factors were described as follows:

A. Depth to Groundwater (D)

The depth to the groundwater in the study area was significantly less than 200 m in all wells (WAJ, 2005). Thus, as shown in **Table 5.16**, the weighting system of the depth index was obtained as a result of multiplying Dr * Dw = 5 (Knox *et al.*, 1993; Aller *et al.*, 1987), where the depth rate = 1 and the weight = 5.

Table 5.15: Rate and weight for the parameters of the DRASTIC Index (Aller *et al.*, 1987; Knox *et al.*, 1993; Piscopo, 2001).

Parameters	Rate (Xr)	Weight (Xw)	XrXw
D: Depth to groundwater (m)	1	5	5
Less than 200 m			
R: Recharge[b] (Range)			
4–6	3	2	6
7–9	5		10
A: Aquifer media (Material)			
Limestone	6	3	18
Basalt	9		27
S: Soil media (Type)			
Silt loam	5	5	25
Clay loam	3		15
T: Topography – Slope (%)			
0–2	10		30
2–6	9	3	27
6–12	5		15
12–18	3		9
I: Impact of the vadose zone		4	
(Material)	6		24
Limestone	9		36
Basalt			
L: Land use			
Urban	8		40
Irrigated field crops	8	5	40
Uncultivated	5		25
{parameter missing}			

All the maps are descriptions for the factors that form the DRASTIC index which was based on the rate and weight (Xr Xw) for each factor included in Table 5.15.

B. Recharge (R)

Recharge to the groundwater is represented as a portion from rainfall amounts. This depends on rainfall data, soil permeability, and the topographic setting by using **Equation 5.3**. In order to calculate the recharge value (Rr*Rw), a digital elevation model (DEM) of the study area was generated for a depth of 10 m. According to **Table 5.15**, the net recharge map was subdivided into two classes, as shown in **Figure 5.7**.

$$\textbf{Recharge Value = Slope\% + Rainfall + Soil permeability} \qquad 5.3$$

Table 5.16: Constructing of the recharge factor (Piscopo, 2001; MoA, 1993; USDA, 1994).

Slope		Rainfall		Permeability		Net Recharge	
Slope (%)	Factor	Rain	Factor	Type of soil	Factor	Range	Rating
< 2	4	< 600	4	Sandy loam	4	3-6	1
2-10	3	600-700	3	Silty clay loam	2	6-7	3
10-32	2	700-860	2	-	-	7-9	6
> 33	1	> 860	1	-	-	-	-

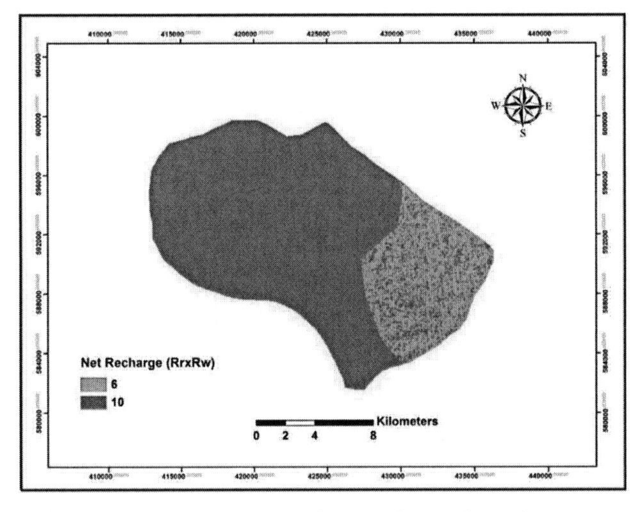

Figure 5.7: Calculated net recharge (RrRw).

C. Aquifer Media (A)

Based on the geological description of the study area, the aquifer media was classified as fractured basalt and limestone **Figure 5.8** and – depending on **Table 5.15** (Knox *et al.*, 1993; Aller et *al.*, 1987) –, the net aquifer media was subdivided into 2. This hydrogeological factor describes the ability of pollutants to move within the aquifer, according to its type.

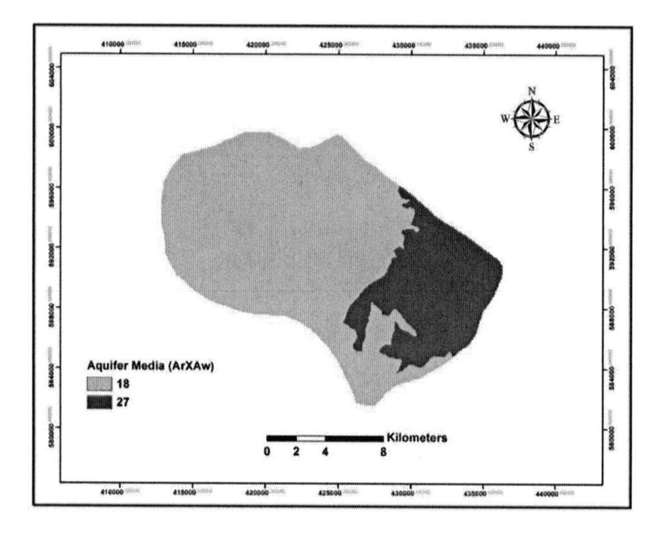

Figure 5.8: Extract of the aquifer (ArAw).

D. Soil Media (S)

Based on the ratings for the soil texture in **Table 5.15** (Knox *et al.,* 1993; Aller *et al.*, 1987), where the soil map was classified into two classes, **Figure 5.9** shows the S factor to building the soil media (Sr*Sw). This factor describes the ability of pollutants to move down to the groundwater within the soil, according to the size distribution of the soil cover.

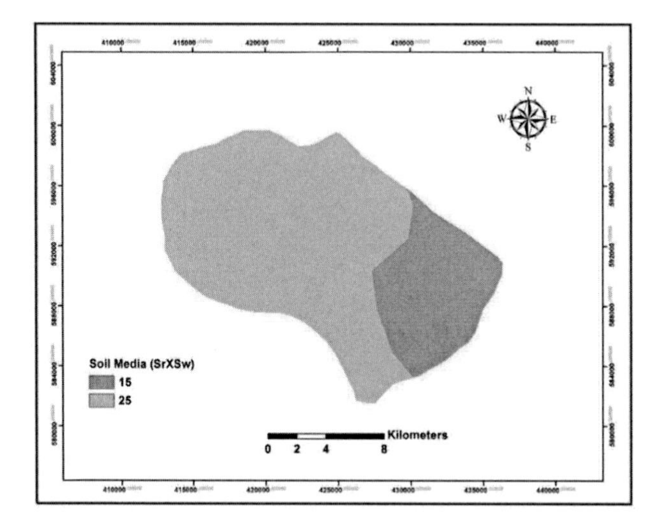

Figure 5.9: Building of the soil media (SrSw).

E. Topography (Slope) (T)

The slope index was derived from the DEM to find the ratings for recharge **Table 5.15**. The topographic slope map was subdivided into five classes and is shown in **Figure 5.10,** where it describes the ability of pollutants to infiltrate into the vadose zone and to reach the aquifer (Aller *et al.*, 1987).

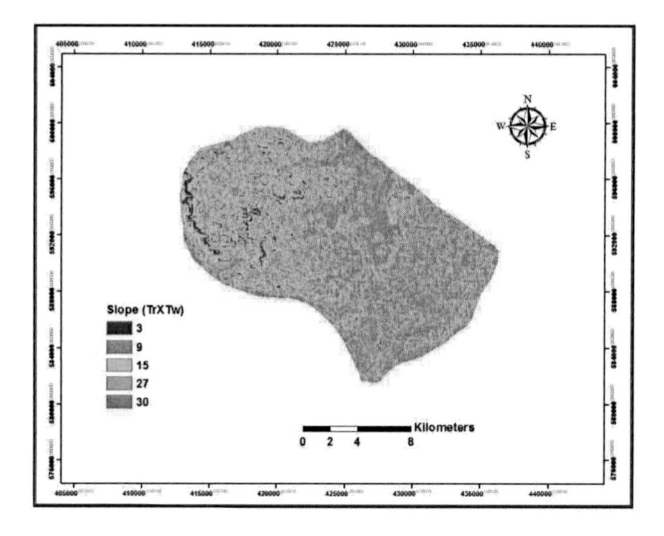

Figure 5.10: Reclassification of the topographic slope (TrTw).

F. Impact of the Vadose Zone (I)

The impact of the vadose zone was examined by interpreting geological maps of the study area; the aquifer media was classified as fractured basalt and limestone. **Figure 5.11** shows map result of impact vadose zone (Ir*Iw) in accordance with Table **5.15.**

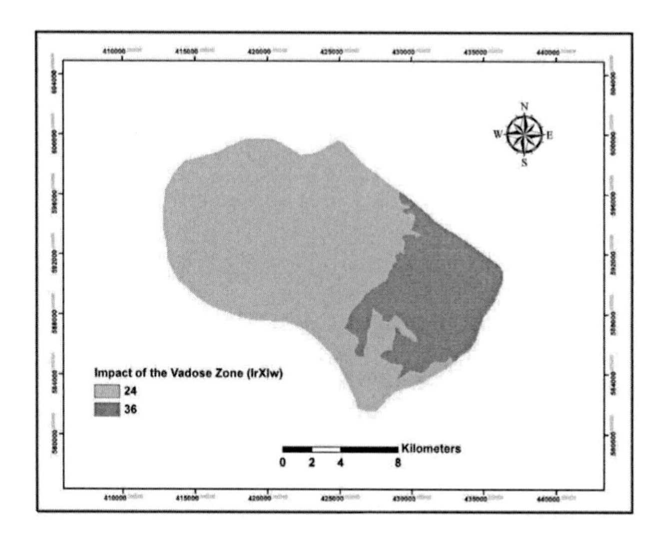

Figure 5.11: Extract of the impact zone (SrSw).

5.6.1.1. The DRASTIC Vulnerability Index

Through the applied DRASTIC index, the final vulnerability map could be obtained for the NE, showing values between 83 and 132. These values were separated into three equal vulnerability classes of low, moderate, and high vulnerability. **Table 5.17** and **Figure 5.12** show the final groundwater vulnerability classes (DRASTIC Index), with of extremely low (1 – 20), low (20 – 40), moderate (40 – 60), high (60 – 80), and extremely high (80 – 100) by dividing the value range into five equal classes according to the standard deviation (SD).

Table 5.17: The DRASTIC index for the study area

Class	Area Km^2	% of the total area
Extremely Low	12,7	4,9
Low	73,1	28,6
Moderate	87	34
High	64	25
Extremely High	19,2	7,5
Total	256	100 %

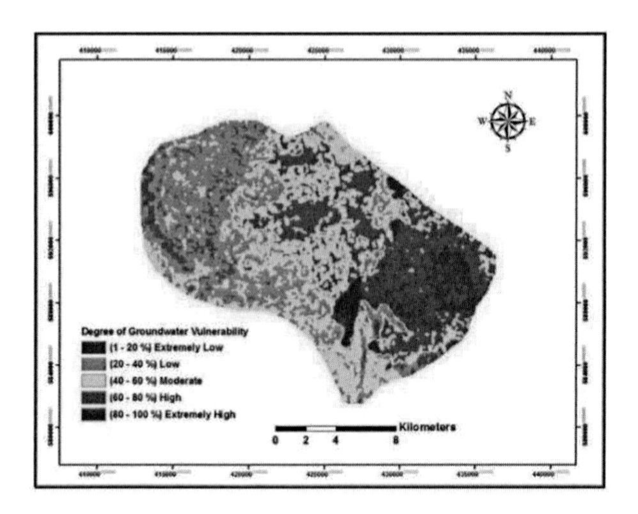

Figure 5.12: Final groundwater vulnerability.

5.6.1.2. Outcomes of the DRASTIC Index Development

The DRASTIC index was modified by adding land-use parameters where land use had to be included in the groundwater vulnerability maps, because different types of land use strongly affect groundwater quality through water infiltration.

In this study, three different land-use classes were defined: irrigated field crops, built-up areas, and uncultivated areas. The areas were manually digitized from the GeoEye-1 sensor and stored in the local coordinate system of Jordan (JTM). Based on **Table 5.15**, the land-use map was subdivided into two classes **Figure 5.13**. The resulting grid coverage was then added to the DRASTIC index.

As it appears from the land-use map, the total used area was 36 km^2, and it was concluded from an overlay that there is a relationship between land use and groundwater vulnerability zones. Farms and urbanization cover 5 km^2 of the extremely high and high vulnerability zones, while about 8 km^2 of farms and urbanization show a low and extremely low vulnerability and the area with moderate vulnerability was 23 km^2; thus it is a possible source of contamination (farms and urbanization), which corresponds with the modified DRASTIC index **Table 5.15** and **Figure 5.14.**

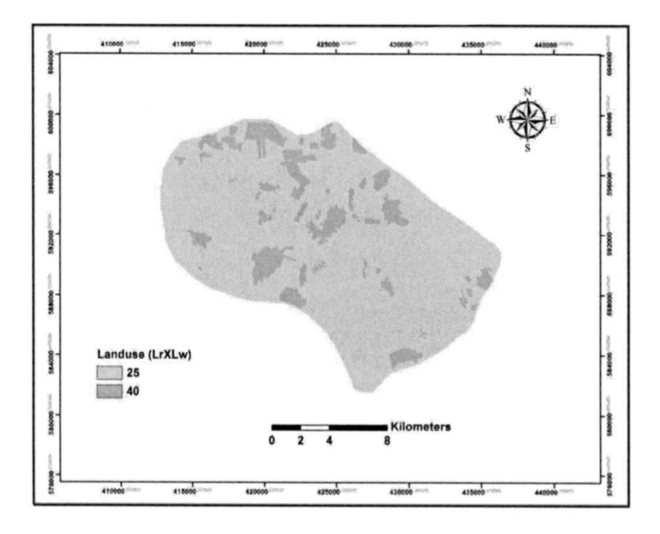

Figure 5.13: Construction map of land Use (LrLw).

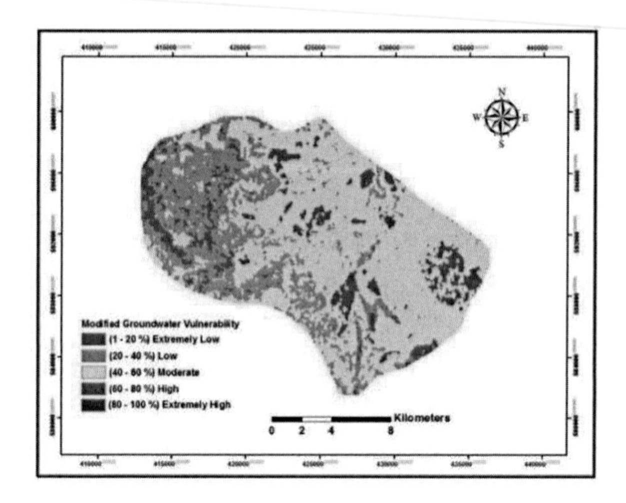

Figure 5.14: Final modified groundwater vulnerability.

Table 5.18 shows the modified groundwater vulnerability classes (modified DRASTIC index) classes of extremely low (1 – 20), low (20 – 40), moderate (40 – 60), high (60 – 80), and extremely high (80 – 100) vulnerability by dividing the value range into five equal classes according to the standard deviation (SD).

Table 5.18: Modified DRASTIC index for the study area.

Class	Area Km²	% of the total area
Extremely Low	**13**	**5**
Low	**72,5**	**28,3**
Moderate	**153,5**	**60**
High	**16**	**6,3**
Extremely High	**1**	**0,5**
Total	**256**	**100 %**

5.6.1.3. Groundwater Vulnerability and Land Use

The spatial relationship between land use and groundwater vulnerability in the study area due to the modified groundwater vulnerability map can be considered with the addition of a land-use map. Through this step, any possible sources of pollution from farm and urban centers can be found that are located within the extremely low, low, moderate, high, and extremely high vulnerability zones. The outcomes of this process are shown in **Table 5.19**, where the relationship between the land-use classes and the groundwater vulnerability zones becomes transparent.

The next table also shows that less than 3% of the study area is located in moderate vulnerability, which it could be a possible source for pollution through farms and urbanization, corresponding with the modified DRASTIC index. The percentage of the study area showing a moderate vulnerability with no possible sources of pollution is 45.1%; on the other hand, about 11.4% have a high and extremely high vulnerability with possible sources of pollution in the same area. The percentage of the study area with high vulnerability and no possible sources of pollution is 9.4%. Around 31.4% have a low vulnerability with no possible sources of pollution.

Table 5.19: The distribution of land use within groundwater vulnerability zones.

	Land-use classes (area (km^2) / % of total area)	
DRASTIC	**Farm / urban center**	**Uncultivated land**
Low and extremely low	-	80 km^2 (31.4%)
Moderate	7 km^2 (2.7%)	115 km^2 (45.1%)
High and extremely high	29 km^2 (11.4%)	24 km^2 (9.4%)
Total	255 km^2 (100%)	

5.6.1.4. Calibration method between groundwater vulnerability and SAR

The quality of water for irrigation reflects inputs from the soil, pollutant sources, and the atmosphere; it also depends on the nature and composition of the soil, sub-soil, climate, topography, etc. (Dhembare, 2012). Thus, irrigation water depends on dissolved salts such as Na, Ca, Mg, and HCO_3, while the ratio of their concentration affects the quality of water for irrigation (USEPA, 1974). Increasing the concentration of salts in irrigation water changes the soil quality (Dhembare, 2012). The sodium absorption ratio (SAR) has been used to evaluate water-quality suitability for irrigation. This is considered to be one of the most common factors that influence the usual rate of water infiltration (Muthanna, 2011).

The objective of the present study was to assess the SAR in wells as an index for the water quality within groundwater vulnerability zones; based on these measurements, the suitability of different uses of the water (i.e. domestic, irrigation, and drinking purposes) were evaluated. The analysis was conducted by using the modified DRASTIC index in combination with GIS techniques and remote sensing. There is a significant relationship between SAR values of irrigation water and the extent to which sodium is absorbed by the soils. Continued use of water with a high SAR value leads to a breakdown in the physical structure of the soil, caused by excess of sodium absorption ratio.

In order to examine the relationship between groundwater vulnerability and SAR concentrations, locations of wells were appointed by using GPS. Water samples from each well were analyzed in the lab. SAR values were distributed throughout the study area in order to implement overlay between wells location and SAR concentration and modified DRASTIC index to assess the situation of groundwater within study area; it was also found that SAR concentrations in wells inside the low-vulnerability zone vary between 10 in well YS to 18 in well AK. **Figure 5.15** shows the spatial distribution of the SAR within the groundwater vulnerability zones.

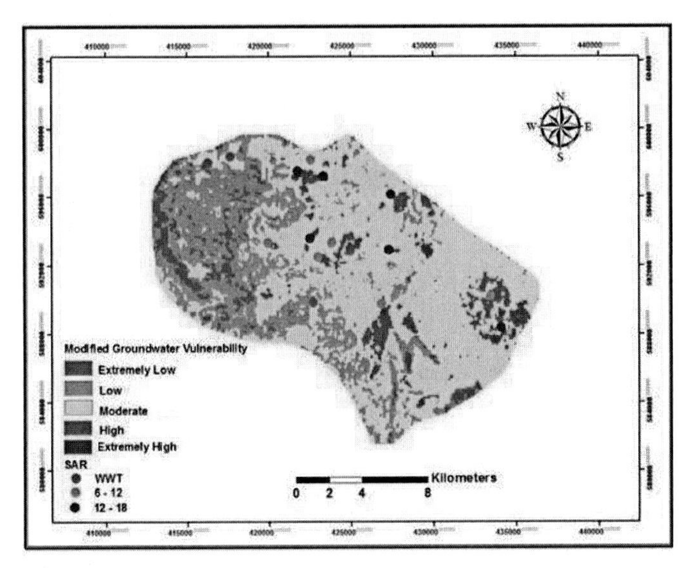

Figure 5.15: Spatial distribution of the SAR in the groundwater vulnerability zones.

All wells inside the groundwater vulnerability zones show SAR concentrations between moderate and very high, according to the international standard for arid and semi-arid areas. A relationship was found between high values of the SAR and the modified DRASTIC index. So the correlation between the DRASTIC values and the SAR were calculated based on Pearson's correlation factor **Table 5.20**. The correlation was 23%, which is relatively low. This means that the intrinsic vulnerability index needs to be modified in order to show a realistic assessment of the pollution potential in the area.

Table 5.20: Correlation factor between SAR and original groundwater vulnerability.

Pearson's correlation	Number data of SAR	Classes in DRASTIC index
20%	3	Low and extremely low
67%	10	Moderate
13%	2	High and extremely high
100%	15	Total SAR

Therefore, land use was added as a modified factor on the DRASTIC index as the source with the highest pollution potential; then, this modification was investigated by assessing the correlation between the DRASTIC values and the SAR, where the highest mean of the SAR was correlated with the highest rate on the DRASTIC index. **Table 5.21** shows the results of this modification for all DRASTIC as well as the land-use parameters.

Table 5.21: Correlation factor between SAR and modified vulnerability index.

Pearson's correlation	Figure in data of SAR	Classes in DRASTIC index
-	-	Low and extremely low
33%	5	Moderate
67%	10	High and extremely high
100%	15	Total SAR

Using the new parameter, a new DRASTIC map was developed, showing that 11.4% of the considered area falls into the high-vulnerability classification. This percentage had been 1.9% before modification. The calculated area was 67% and 20% for the moderate- and the Low and extremely low classes, respectively, and 13% for the High and extremely high classes. There was no percentage for the low class, but 33% for the moderate and 67% for the high ones after application of the modification. These results show a clear effect of the modification. In addition, **Figure 5.15** shows the spatial distribution of the index after the modification.

5.6.2. The Results on the Development of Land Degradation

The final step comprises matching the physical environment qualities (soil, climate, and vegetation indicator) and the management indicator for the definition of the various types of LDD. The four derived indicators are collected for the assessment of the LDD index (LDDI) in **Equation 5.4**:

$$LDD = (SI + VI + CI + MI) \quad \text{Equation 5.4}$$

The LDD is defined on a three-point scale ranging from a high value (high sensitivity) to a low value (low sensitivity), for a better integration of the boundaries of the successive LDD classes. This methodology was validated in the study area by using geoinformatics techniques, such as the correlation factor between pixel values of bands in remote-sensing data and change detection, as well as field studies of the physical and chemical properties of the soil.

Based on the results of fieldwork and data collection for each indicator, a model of the degradation, the mapping of indicators, and the performance of mathematical equations can be applied. The following equations apply to each indicator with parameters:

Soil Indicator

This indicator has five parameters (parent material, soil texture, depth of soil, slope, and organic matter) which will be explained, and the score for each parameter will be extracted in order to make the soil calculations. The parent material was classified as fractured basalt with values reaching 5 and limestone with values reaching 15. In accordance with **Table 4.9**, the parent material was subdivided into two classes (Kosmas, 1999; Contador *et al.*, 2009). **Figure 5.16** shows the mapping through the parent-material parameter (Pr*Pw).

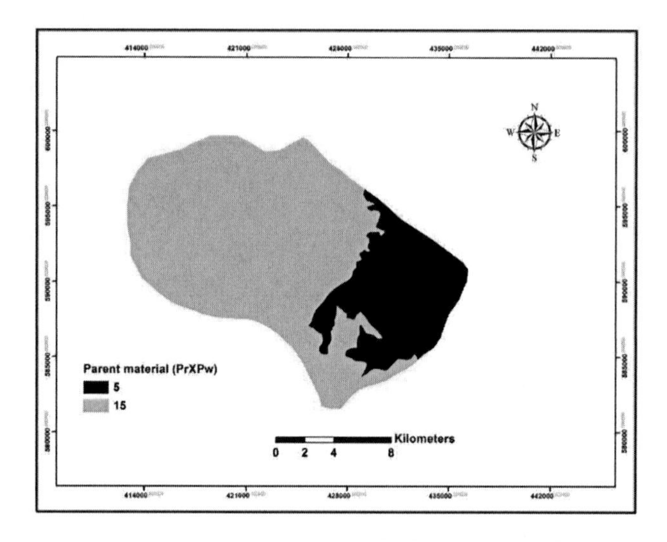

Figure 5.16: Construction of parent material (PrPw).

The soil texture was classified into one class, based on its ranking **Table 4.10** (USDA, 1987; FAO, 2006). It was obtained as a result of multiplying Tr*Tw = 11.65, based on the weighting system of soil **Table 4.23**; its significant depth in the study area was about 140 cm (MoA, 1993). The depth index was obtained as a result of multiplying Dr*Dw = 5, based on the weighting system in **Table 4.11**. And the slope index was derived from the DEM, based on **Table 4.12**; the topographic slope map, sub-divided into three classes, is shown in **Figure 5.17** (Sr*Sw).

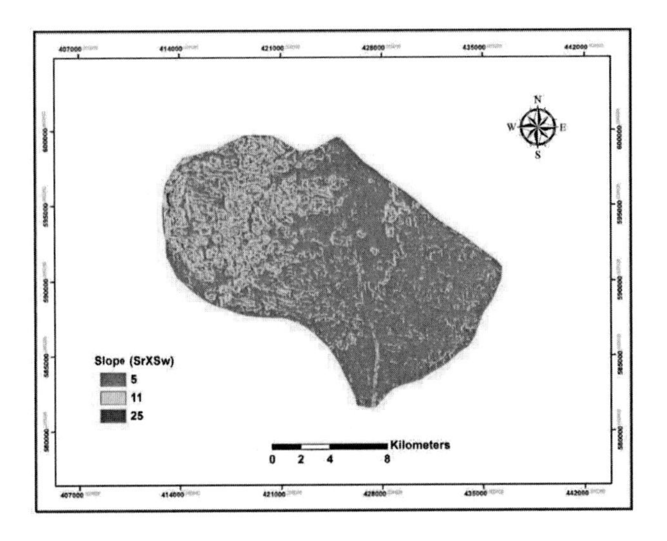

Figure 5.17: Reclassification of slope (topographic, SrSw).

On the other hand, the organic matter index was derived from the sample analysis (fieldwork in 2013) and is based on **Table 4.13**; the organic matter results were sub-divided into two classes; 20 and 25; this is shown in **Figure 5.18** (Or*Ow).

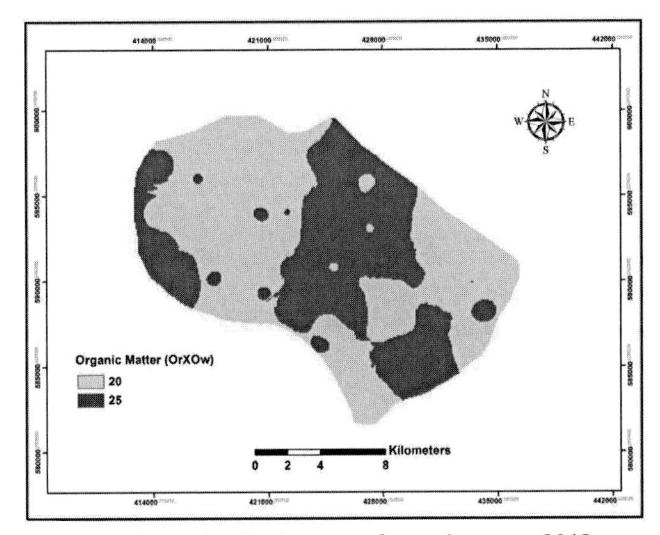

Figure 5.18: Distribution map of organic matter 2013

Vegetation Indicator

This indicator has three parameters: vegetation land, vegetation cover, and devegetation, which will be explained and their score parameters extracted in order to calculate vegetation. The vegetation land map was classified into two classes: citrus, conifers, deciduous and olives trees, with values reaching up to 6; and annual agricultural crops, green and barren land, with values reaching up to 15, based on the ratings for the types of vegetation land **Table 4.14**, which are shown in **Figure 5.19**. The percentage of the vegetation cover has been measured with satellite images; depending on these results and based on the ratings of the vegetation cover, **Table 4.15** shows that the Vr*Vw = 9. For the percentage of removable vegetation, that depends on the ratio of annual vegetation losses on the total surface area. In this factor, techniques of remote sensing have been used in order to measure percentages for the removal of vegetation where the annual loss of vegetation was at about 1.6%, and in accordance with Table 4.16, the RVr*RVw = 9.

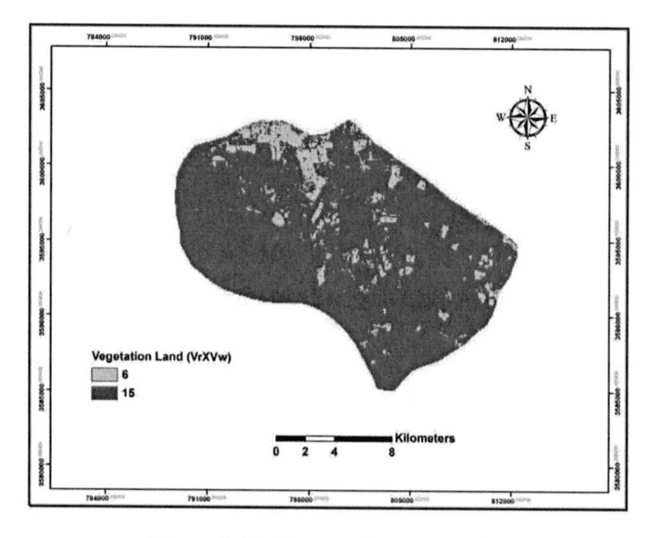

Figure 5.19: Classes of vegetation land.

Climate Indicator

This indicator has two parameters (aridity and rainfall), which will be explained, as well as the score extracted for each parameter in order to make the climate calculation. The aridity parameter depends on the aridity index that includes the mean annual precipitation, while MAE is the mean annual evapotranspiration. Based on **Table 4.17**, the ranking of the aridity parameter is 4, which means the Ar*Aw = 12. While the rainfall rate depends on the annual rainfall being distributed on regions and is based on the annual rainfall map of Jordan, the precipitation does not exceed 250 mm. Thus based on **Table 4.18**, the score of Rr*Rw = 15.

Management of Land Use Indicator

This indicator has four parameters (density of land use, reclamation areas, water quality index and tillage), which will be explained, as well as the score extracted for each parameter in order to make the climate calculation. The density of land use was distributed into three classes: low, medium, and high. Based on the fieldwork and on interviews with the farmers on the farms that were selected, it was shown that most of the area features a low density of land use. Based on **Table 4.19**, the score of the density of land use has one class (medium), which means the LDr*LDw = 12. While the factor for the reclamation of affected areas depends on the modified degradation index in **Equation 4.9** and is based on **Table 4.20,** the reclamation score of affected areas is shown in **Figure 5.20.**

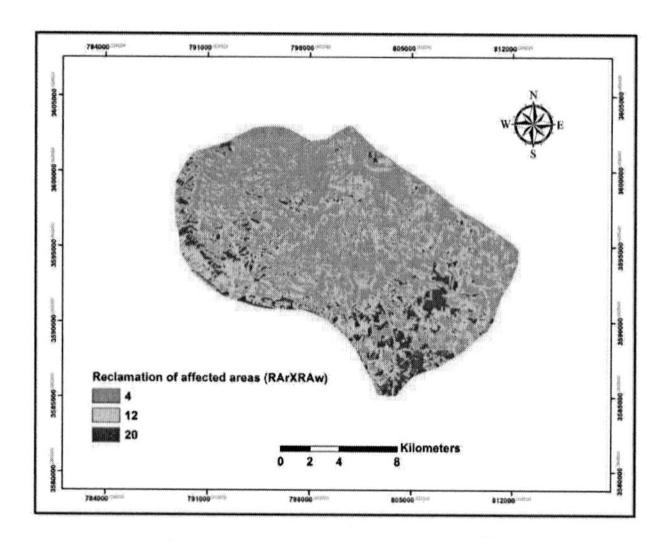

Figure 5.20: Classes of reclamation factors.

For the water quality index which depends on results from the analysis of water wells, water quality values represented by SAR have been extracted; based on **Table 4.21**, the ranking of the SAR in the study area is 5, which means the Wr*Ww = 20. On the other hand – and based on the interviews and fieldwork –, the tillage factor does not cover the entire study area and was limited only to the farms and grassland areas. Based on the results of the classification of the land-use map in **Figure 5.15** and **Table 4.22**, the tillage in the study area is shown in **Figure 5.21.**

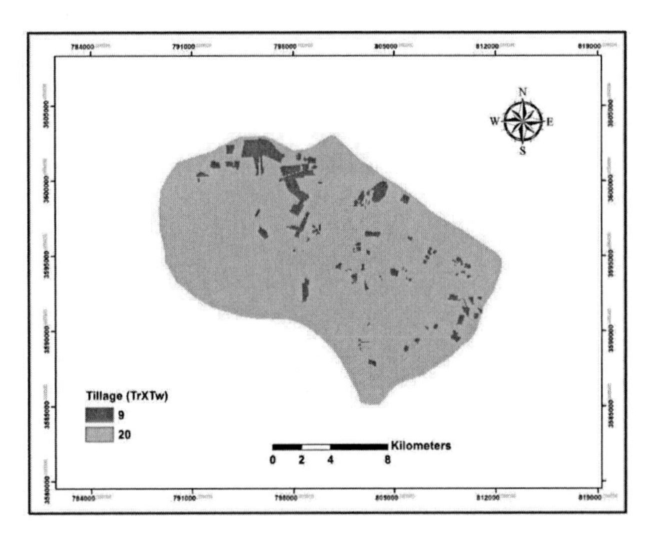

Figure 5.21: Classes of tillage factors.

5.6.2.1. The Degradation Index

The final degradation degree map for the NE showed values between 142 and 210, which were separated into vulnerability classes of low, moderate, and high vulnerability by dividing the value range into three natural classes. **Table 5.22** shows final degradation classes with (a) 142–170 (Low), (b) 170–181 (Moderate), and (c) 180–210 (High) in **Figure 5.22.** It was found that the largest portion of the area is moderate, reaching 45% **Table 5.22**, while low degradation covered 21% and high degradation reached 34% in the study area.

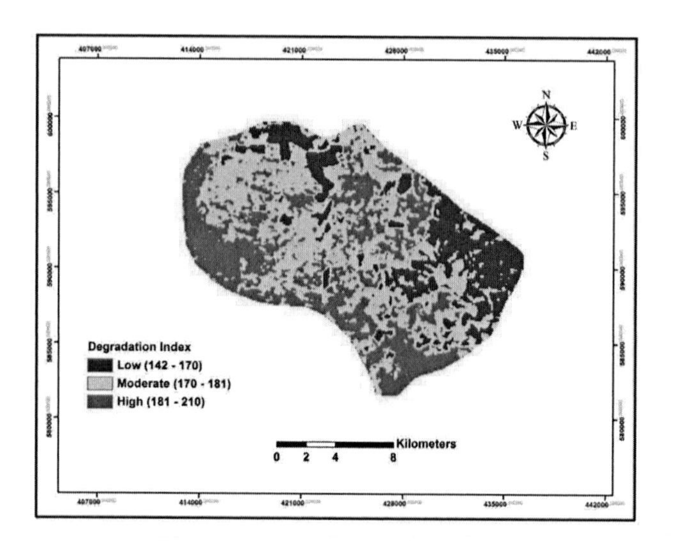

Figure 5.22: Classes of the degradation index.

Table 5.22: Degradation index for the study area.

Classes	Degradation index	Area in km^2	% of the total area
Low	142 - 170	54	21
Moderate	170 - 181	115	45
High	181 - 210	86	34
Total	-	255	100

5.6.2.2. Validation of Results

In order to assess land degradation, the model applied two methods, which were carried out through fieldwork and remote-sensing techniques as follows:

I. Field studies

1. Soil structure

During fieldwork in 2013, it appeared that the study area had two soil structures, and both were arid and semi-arid areas, which indicates that the soil is weak or moderately weak **Section 5.3.1.1, Part D**.

2. Soil horizon

From fieldwork data in 2013, it appeared that there was one type of soil horizon covering the study area (Ap), which means that the area lacks characteristic properties of soil such as E or B horizons or that this property is a result of human practices in agriculture or similar disturbances, which confirms the deterioration in the soil **Section 5.3.1.1, Part E**.

3. Sodicity

The results of sodicity appear in **Section 5.3.1.2, part C**. Most samples with high values (> 8) were located in low and moderate areas on the land degradation map and the rest of the samples were located.

II. Remote-sensing data

Correlation factor

Spatial analysis tools in remote sensing techniques and bands of satellite imagery were used in order to find the relationships between the land degradation model and other factors:

1. Correlation factor between land degradation model and salinity index

In order to extract the correlation factor for this relation, the following equation was used:

Salinity index = (B4-B5)/(B4+B5) **5.5.**

Where the B4 is near infrared and B5 is short-wave infrared in the Landsat TM, **Figure 5.23** shows the salinity index for the study area.

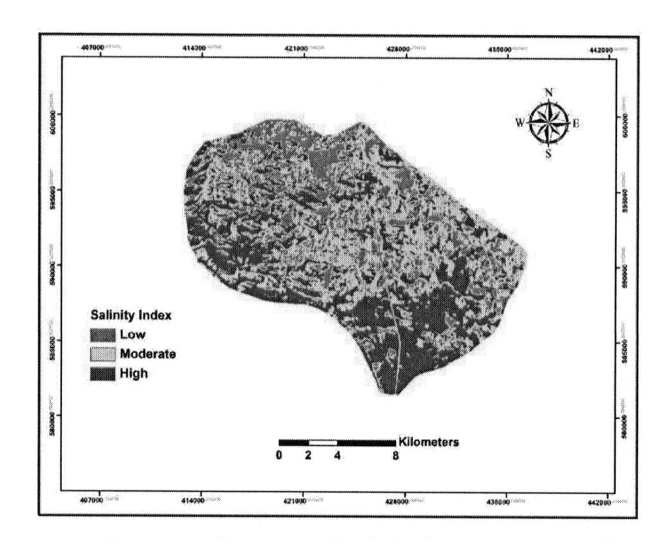

Figure 5.23: Salinity index.

The results of the correlation factor indicate that the R^2 between the land degradation model and salinity equal 0.7168 (**Figure 5.24**).

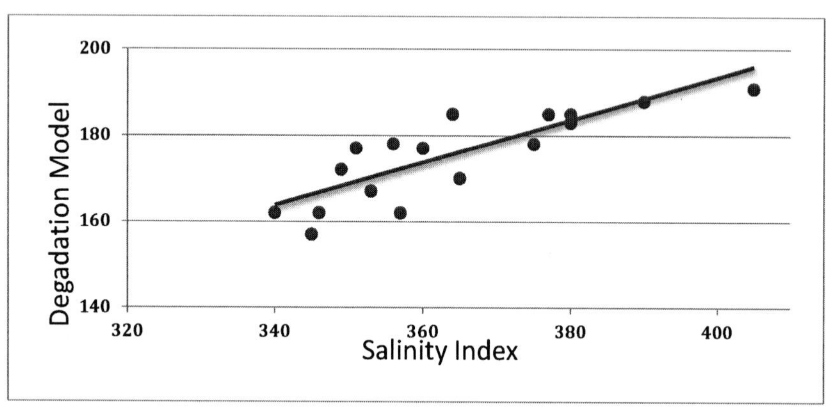

Figure 5.24: Correlation between the land degradation model and the salinity index.

2. Correlation factor between the land degradation model and the degradation index

In order to extract the correlation factor for this relation, the calculation was conducted with **Equation 5.6**:

$$\textbf{Degradation index} = 255 - (B2+B3)/255 + (B2+B3) \qquad \textbf{5.6.}$$

Where the B2 is near infrared and B3 is short-wave infrared in the Landsat TM, **Figure 5.25** shows the degradation index for the study area.

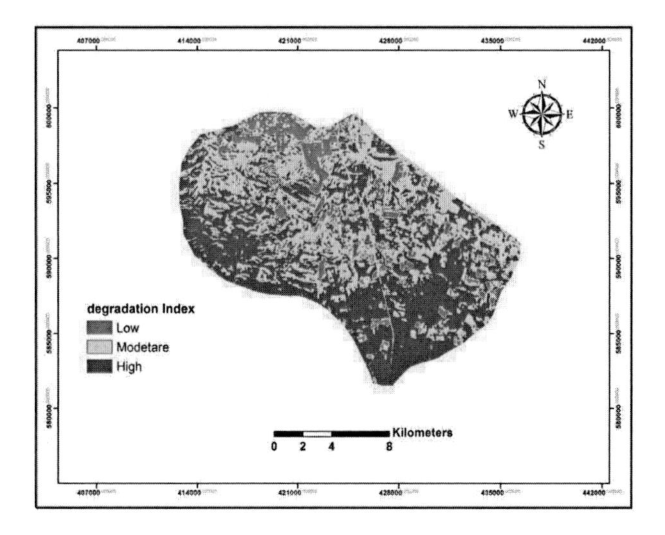

Figure 25: Degradation index.

The results of the correlation factor show that the R^2 between the land degradation model and salinity equal 0.7283 **Figure 5.26**.

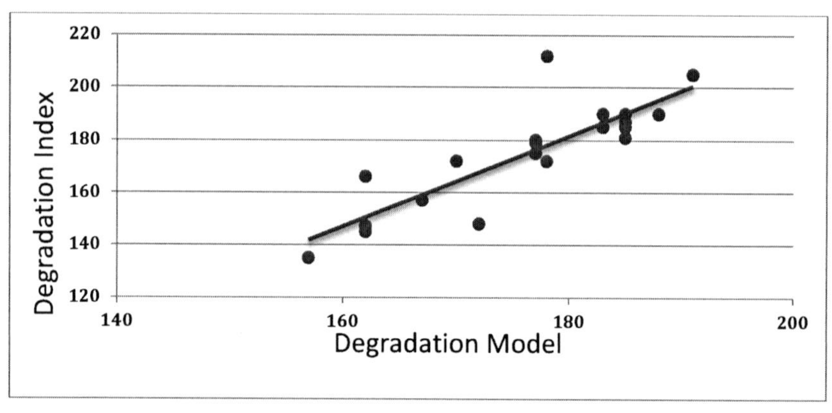

Figure 5.26: Correlation between the land degradation model and the degradation index.

There is a significant correlation in the previous relations where the statistics of this analysis is shown in **Table 5.23.**

Table 5.23: Significant correlation for the relations.

	P Value	Multiple R^2	Adjusted R^2
Correlation 1	6.78e-07	0.7168	0.727
Correlation 2	4.454e-07	0.7283	0.7147

Results of these methods show that geoinformatics techniques could be used as assessment tools in the resulting model, which could lead to a more realistic management of land degradation.

3. Relationship between the land degradation model and SAR values

Figures 5.27 A, B, and **C** show the relationships between the pixel values of the land degradation model and SAR values in different areas where a significant correlation between the land degradation model and SAR values ($p = 0.000$, $p = 0.00$, and $P = 000$) can be found for the areas Protected, Cultivated and Abandoned, respectively. For protected, cultivated and abandoned lands, R^2 was 0.6406, 0.675, and 0.725, respectively **Table 5.24.**

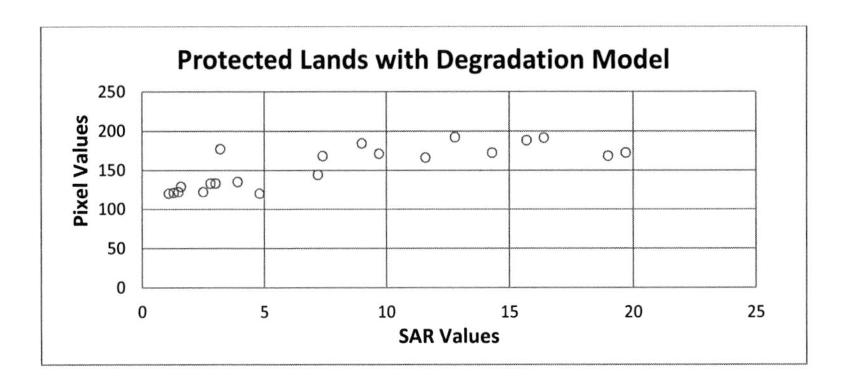

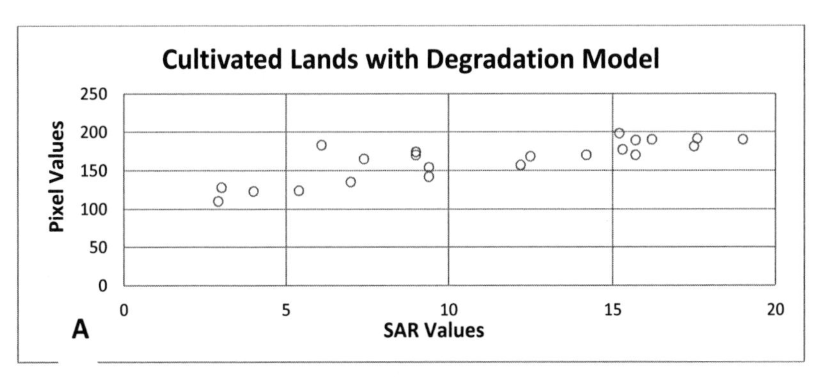

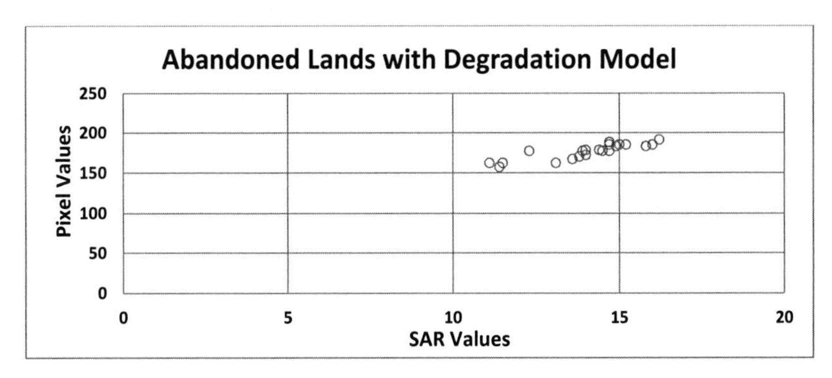

Figures 5.27: Relationships between the land degradation model and SAR values in the different lands: A) SAR values in protected lands with pixel values of the degradation model, B) SAR values in cultivated lands with pixel values of the degradation model.

Table 5.24: Statistical values for the relationships between the land degradation model and other factors.

	P Value	Multiple R^2	Adjusted R^2
Relationship 1	7.717e-06	0.6406	0.6227
Relationship 2	2.395e-06	0.675	0.7113
Relationship 3	5.025e-07	0.725	0.7113

5.7. Hot-Spot Areas between the Vulnerability of Groundwater and Land Degradation

In order to evaluate the relationship between the vulnerability of groundwater and land degradation, intersecting points must be investigated representing hot-spot areas in the region that are at high risk. With the results of the modified DRASTIC index and the land degradation models, a map was extracted that reflected hot-spot areas in the region, as shown in **Figure 5.28**. Most of the hot-spot areas were located in the center and north-east of the study area; the total area of hot spots is about 7.2 km², which constitutes a threat for the entire area, especially due to the active human and agricultural practices.

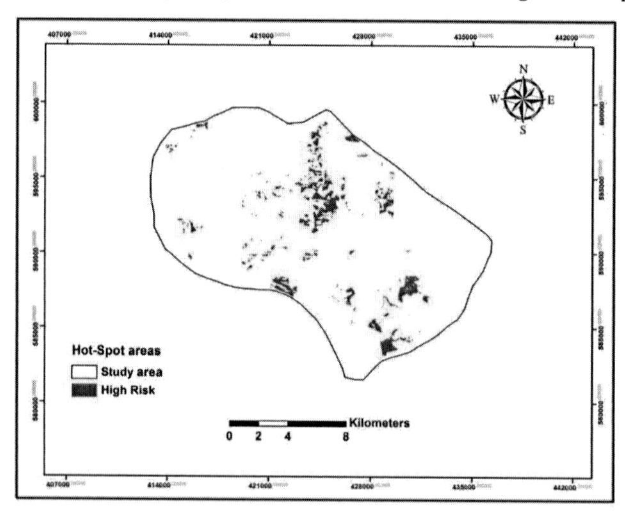

Figure 5.28: Hot-Spot areas between the vulnerability of groundwater and land degradation.

5.8. Strategies of Risk Management

5.8.1. Groundwater Vulnerability

I. Assess management of the risk

The vegetable farmers are continuously moving their farms within the study area from one location to another, and with the estimated degree of salinization through sodium absorption ratio (SAR) where infiltrate to groundwater that used to irrigation, this constitutes a main risk to groundwater quality in the aquifer of the Yarmouk basin. In addition, the nutrient load from both pollution sources (cultivation of land and water from wells) has an impact. In order to overcome the problem of land salinization in the study area, some locally acceptable solutions are suggested in the recommendations. In the event of failure to take any preventive action to control the increase of salinization in the area, it could cause groundwater pollution and further land degradation. To do nothing about this situation could possibly put this risk into reality in the surrounding areas.

II. "Management strategies"

It was found that on each farm, farmers had used more fertilizers than required for plant uptake, especially inorganic ones. Even other problems are based on the use of fertilizers; Al-Adamat (2002) suggested the fertilizer management is necessary to tackle the salinity problem. It is possible to calculate the amount of inorganic fertilizers used in cultivation and to replace them by bio-fertilizers within the standards for inorganic and organic material used in agriculture, as well as to monitor the pumping of water from wells in order to control the water level in the study area.

5.8.2. Land Degradation

I. Assess management of the risk

The degradation issues existing in the region focus on two sides. For the larger part, the responsibility is on the farmers through farming practices and human management of agriculture, such as the control of irrigation water usage, the quality of fertilizers, ploughing mechanisms, reclamation of land within the boundaries of agricultural as well as non-agricultural land, and pressure control of soil salinity, which constitutes the main risk to soil and groundwater quality. In addition, nutrient which represent other pollution sources through excessive increase in the use of fertilizers and pesticides (the quality and quantity of fertilizers). In the event of failure to take any preventive action to decrease salinization in the area, the degradation of lands could increase in the study area. To do nothing about this situation could possibly increase this risk in the surrounding areas and their natural resources. The other partner in this is the government. It is their responsibility to adjust some preventive procedures and to be serious about the implementation of laws in addition to some of the procedures mentioned in the recommendations accepted locally to reduce land degradation.

II. "Management strategies"

It was found that each farmer has a special policy in sustainability and reclamation considering what is going on throughout the region in terms of climatic conditions. There are suggestions that serve the interests of agricultural land: adjust agricultural management and the awareness of workers on farms; control water pumps to reduce the salinity in the region; increase reliance on bio-fertilizers in order to increase soil fertility; and controlled usage may be possible for inorganic fertilizers.

The agricultural and economic situation of the farmers in Jordan is poor and being aggravated by relatively small farm sizes, human activities. Different factors limit agricultural production and this also lead to land degradation. There are two perspectives included in the proposed management plans could be taken into consideration: measures to be performed by the farmers themselves and the combinations between the management

5.8.3. Proposed Management Plan for the Farmers

1. Addition of manure to the soil according to limited standards.
2. The need to use micro-elements as fertilizers because some farmers only use macro-elements in balanced proportions.
3. Sources and species of seed should be the same. Planting different species on one plot will reduce crop yield.
4. It is important for farmers to use the agricultural guides for each crop of the Ministry of Agriculture rather than simply relying on their experiences in

agricultural production. They should apply the agricultural management processes as stipulated in the management guidelines.

5. Planting the same crop and not different ones in one area.
6. Monitoring the level of groundwater to prevent over-pumping.

The results from the questionnaires indicate that the farmers need instruction and learning guidelines. Therefore, it is very important to teach them how they can apply the new technologies in agricultural production in order to decrease yield losses, to increase the quality of management processes (weed control, spraying, irrigation, etc.), to save time, to decrease the costs of agricultural production, and to increase yield production. Also, the government should provide farmers with basic farming requirements (fertilizers, seeds, and pesticides) at reasonable (suitable) prices. As yet, they are unable to purchase these items on the free market.

5.8.4. Proposed Management Plan for the Government

Plans of this kind need management plans, which are very expensive to implement. They also require great expertise and experience in agricultural management:

1. Monitor the quality of the soil by collecting samples each year to recommend the right chemical fertilizers for agricultural production.
2. Manage agricultural lands in accordance with the crops or yield in order to mitigate the pressure on lands which lead to mitigate land deterioration.
3. Map agricultural lands by using geoinformatics or high-resolution aerial photographs; use detailed knowledge of soil types and GIS.
4. Analyze manure in order to determine the nutrients it provides for plants.
5. The agricultural growth model is a useful tool to identify areas that have similar soil and plant parameters (yields, variable nutrients, organic matter, pH, etc.).
6. Seriously try on the government's part to find realistic solutions for the water problem and to control weakness points in groundwater zones.

5.8.5. Managing Soil Salinity

Where soils have a high level of salt, the solution is to leach it and replace the sodium with calcium where changing soil salinity is the most difficult option. Usually, there are some other options for correcting these soil-related problems:

1. Change the salinity and alkalinity of the soil.
2. Use the plant species which are more tolerant to salinity, as well as biotechnology in order to address problems faced on agricultural lands.

The methods commonly adopted or proposed to reclaim saline soils include salt leaching and (surface and subsurface) drainage where these methods are used to remove soluble salts from the root zone (FAO, 1988).

5.8.6. Managing the Sodicity of Irrigation Water

Sometimes, poor quality water is used for the irrigation of agricultural lands and it necessary to avoid soil problems, because these problems could impact on soil quality and this lead to the degradation of agricultural lands, where these problems are related with practices of using water..

1. Decrease the sodium-absorption ratio (SAR) to a safe value by adding soluble calcium such as gypsum (calcium sulfate). If dispersed it should be applied directly ahead of irrigation or thoroughly incorporated into the tillage layer to avoid crusting problems. If gypsum is used, the leaching requirements may be increased.

2. Increase adequate internal drainage if barriers restrict the movement of water through the root zone.

3. Used fresh and drainage water and meet the necessary leaching requirement of water (over-irrigation) if it is available and enough. This is necessary in order to avoid the accumulation of salt in the soil solution.

4. High levels of available water should be maintained in the soil. The soil should not be allowed to get more than moderately dry, since the crop cannot remove all the normally available water due to the higher salt content. Salt and sodium should be monitored with saline-alkaline soil tests every one to two years to extract the SAR saturation or a percentage of exchangeable sodium in order to detect changes before permanent damages occur.

Chapter 6

Conclusions and Recommendations

6.1. Introduction

This chapter aims at critically suggesting possible improvements for subsequent studies that apply satellite imagery and are GIS-supported by fieldwork to monitor land degradation for agricultural management in Jordan and other regions that face the same problem.

The study area was located in the northwestern part of Jordan and was selected due to four factors: the availability of satellite images for the region (past and present); the availability of previous scientific studies carried out in the region and consequently serving as a source of reference or ancillary information; the area features land use and cover classes representative of the whole of Jordan (agriculture, bare soil, salt, and urban areas); and lastly, there are many variations in soils types, which affects yield production.

Impact-farming practices are very prominent in this study area and are responsible for the degradation of some agriculture land. The whole study area was used to monitor changes in land use and coverage – either increase or decrease in agricultural area – by using satellite images from two different periods.

Soil samplings and water were selected from the whole study area, representing three types of land, in order to investigate the human impact on agricultural land, depending on the variations in the management processes and soil types.

Agricultural land in the study area is dependent on irrigation. The hot and dry climates of these regions require that irrigation water does not contain soluble salts in amounts harmful to the plants or having an adverse effect on soil properties. Water of such quality is usually not available in sufficient quantities to satisfy the requirements of all crops grown there. Under these conditions, the farmers are obliged to use water frequently, leading to over-pumping and the use of irrigation water with high quantities of dissolved salts, invariably accompanied by yield losses of most crops. The indiscriminate use of such water can often lead to crop failures and to the development of saline or sodic soils, which in turn require expensive treatment to make them productive again. Water plays an important role in soil-forming processes, especially the groundwater table (its depth and salinity), which is considered the decisive factor of salinization, since it affects the salt balance of the soil and directly influences the exchangeable sites in the soil complex.

6.2. Limitations and Recommendations

The use of geoinformatics technologies in Jordan is still generally in its initial stages, and apparently most policymakers remain relatively unaware of its benefits. Consequently, the achievements to date may not be spectacular compared to the strides made in the world in general. Due to irregularities characterizing the agricultural and water sectors, where conditions are still uneasy and the work atmospheres characterize the use of remote-sensing and GIS techniques. The major limitations in this research project could be grouped into two categories: (a) data limitations and (b) methodological limitations.

Remote sensing missed annual data which has formed constraints to assess the changes in land use by the agricultural activities in the last two decades in the area. The spatial resolution of the remotely sensed data used in this research could be improved by using higher-resolution imageries such as SPOT and IKNOS. They could be used to map and quantify more accurately the land-use change in the area.

Still, historical data about groundwater hydrochemistry was not available for all wells in the area – especially the (public and private) irrigation wells –, which prevented a better understanding of the temporal variations in the area's groundwater quality. This made it difficult to compare between the water quality variation in the irrigation and drinking-water wells, and increases the difficulty experienced with the stage wells in the event of a problem.

Lack of understanding of the research and awareness of the people and usefulness of these research projects formed another barrier to completing most research, especially during the interviews, where most of the samples were located within the private farms, which poses a difficulty for collecting them or requires license. Sampling soil from cultivated land in 2012 and 2013 created problems with some farms.

The conduct of proceedings were weak to access to databases and information, where there are no rules to determine the standards and protocols of data collection, storage, labeling, and integration, where the most urgent and frequently occurring issue is the need to define strategic objectives for enhancing access to information and its sharing. Access to geospatial information is very limited; the results of GIS and remote sensing analyses including databases, maps, and studies are inexistent and access is very restricted in cases where they do exist.

The government carries out severe pressure on farmers through the taxes on water and agricultural yields, these procedures make farmers reticent on necessary agricultural information that would help complete the research.

The DRASTIC model depends on the simple mathematical relation in accordance with expert opinions, although the model is widely used to assess groundwater vulnerability and has little emphasis on the processes that control groundwater contamination. Through this model could be advisable to increase studies of the region in different ways in addition to qualitative studies on water quality, soil, used fertilizers, and linking them with results of water analysis as well as monitoring water wells and contaminants, because in most often the surface can affect the quality of the water or soil quality water bearing.

For the land degradation model also depends on simple mathematical and is motivated by data availability through a number of indicators, including a number of parameters. It has the flexibility to focus on the sources of deterioration through increasing the number of transactions in the event of the availability of the necessary data.

In order to reduce the impact of agricultural practices should monitor of these practices like over-pumping of water, activate reclamation land, correctly use of fertilizers, activating the agricultural extension and awareness of the workers. It is highly recommended that research be carried out in the area to set general guidelines for agricultural practices, including the recommended type and amount of fertilizers and pesticides that give farmers the required crop yield with a minimum risk of contaminating the soil and groundwater.

The future research focus on the changes of land use in different areas, where these researches are needed to understand and explain the problems associated with the spread of agricultural practices and impact of potential environmental on the changes of agricultural. On the other hand, support is needed for geoinformatics techniques through briefings and workshops where all policymakers and stakeholders have an opportunity to gain a better understanding of the maps and analyses that are being developed and presented by experts, as well as their application in the study of natural resources.

Establish a soil and wells database include soil quality, groundwater depth, hydrochemistry analysis for irrigation and drinking water. Both decision-makers and researchers could use such a database to assist in managing groundwater resources and to study groundwater quality. In addition, future workshops, technical exchanges, and collaborative research activities are necessary to strengthen the national capacity for monitoring natural resources and environmental degradation, with a particular focus on soil. The latter is particularly important. Given the paucity of technical expertise in the region.

Finally, it is also necessary to improve the awareness and access of available satellite imagery and spatial information to local remote-sensing technicians.

6.3. Conclusions

This research confirmed the possibility of integrating several parameters to build a land-degradation model by using geoinformatics to investigate the impact of human practices on the degradation of soil, where the quality of results for the land-degradation model depends on the quality of parameters and their source data. Many parameters were obtained with reasonable sources and accuracy by using remote-sensing techniques such as the degradation index, the salinity index, land use and coverage, interviews with farmers, field surveys, and lab analyses. The accurate representation of surface samples in the study area and the mapping of land degradation related to soil properties were implemented on a regular basis. This research also presented a methodology for the prediction of the degradation degree relevant to agricultural practices and soil properties, based on the impact of human activities on agricultural lands. The objective and focus of this study was not just to construct the land-degradation model, but to show the impact of farming practices on agricultural lands by using several parameters characterized by simple methods, effective manner, and, at the same time, affordability.

The analysis of parameters showed that one parameter sometimes has one value covering the entire study area. Also, the lab analyses showed the impact of agricultural practices through fertilizer types and measurements of soil quality, the quality of water used for irrigation, and the disparity in physical and chemical properties of the soil. The analytical results also showed a rise in salinity in the abandoned and some cultivated lands, in addition to rising salinity in the water wells that exceed the limit of what is permissible. On the other hand, the geographic information systems (GIS) provided a suitable environment for modeling the results of these analyses and were used within the parameters of the model used in this research.

With regard to groundwater vulnerability, it was found that developing a DRASTIC index by adding land-use parameters increases the effectiveness of the DRASTIC index to highlight the areas most susceptible to contamination. Especially since the coefficient of land use shows areas that are likely to coincide with sources of pollution more than others, it was found that using the SAR as a correlation factor could be efficiently utilized to show a realistic assessment of the pollution potential in the area before and after adding the land-use parameter to the DRASTIC index.

Four major factor related with land degradation include soil, vegetation, climate, and management of lands; these factors comprise fourteen parameters, parameters were selected in accordance with the nature of the region and available data that could be efficiently utilized to evaluate the impact on land degradation. Parameters were built in a way that they are simple and effective to apply, so that the model could map the degree of land degradation degree through them. Due to the existence of a parameter description, open source availability, little processing costs and time requirements, MAP and MAE were considered to be most suitable for deriving the aridity and – like the RS data – the reclamation parameter. The parameters were collected through the algorithm (LDD), which was developed by using weight and ranking. For the degradation of soil, it was found that field-based soil description of one soil profile per square kilometer could provide a large and sufficient database that will help improve the description of degradation-relevant soil properties of the sand and silt, clay, and organic content, as well as sodicity. Field data for the soil were collected by using a systematic method covering the entire study area. It was found that there is a relationship between land degradation and the vulnerability of groundwater through hot-spot areas.

In the validation stage of land degradation, it was found that there is a relationship between land degradation in this model and the salinity index (B2 and B3) that could be applied within a GIS environment by using the RS data, and also the results were shown that there correlation between land degradation model and the degradation index (B4 and B5) with significant values.

Finally, the results of this research were shown that there are effectively methods could be applicable and experimental to measure the land degradation degree in the arid and semi-arid areas like study area in this research by developing and improving the parameterizations of soil, vegetation, climate, and land management. In order to make the results of the land-degradation model more effective and efficient, it has to use high-resolution satellite images in order to provide the largest possible database of land-use data.

A Summary of Principal Research Findings

The study area was selected in accordance with criteria where indicated the presence to farming practices lead to land degradation and potential sources of pollution, , the availability of secondary data, and the accessibility of the study area, to implement this research work was identified sources of land degradation and water pollution that impact on soil or lands like agriculture activities by tillage, use of fertilizers, pesticides that impact on organic matter, irrigation and salinity, in other hand the fertilizers, pesticides and sodicity could impact on groundwater quality

Secondary data included rainfall, topography, hydrology, geology, soil, well data, and several satellite images (**Chapter 3**). Fieldwork was carried out in 2012 and 2013 in order to collect the necessary primary data which included: (a) types of fertilizers and pesticides used in the study area; (b) water samples from both irrigation- and drinking-water wells in order to find out their major cations and anions as well as SAR concentrations; and (c) soil samples in order to investigate the soil quality (chemistry, physical and biological properties) in the study area (**Chapter 4**). An attempt was also made to map changes in land use in the study area by using various satellite images over the period 1990 to 2010, as well as Landsat TM and Geo-eye 1 in 2012 in order to investigate the spatial changes in the study area due to various agricultural activities (**Chapter 4**). The methodology used for analyzing these satellite images was based on the use of the principal component analysis (PCA) (satellite images only) and the digitizing tool in Arc GIS and ERDAS. It was found that the cultivated area decreased from 53 km^2 in 1990 to 22 km^2 in 2010 (both figures based on Landsat TM); all these data and methodologies were used to achieve the objectives of this research.

The First Objective

There was an attempt to map soil degradation by geoinformatics and building a model through indicators that have rank and weight, where the weight depend on the two scenarios (scenarios were developed by using questionnaires were handed out to all stakeholders, farmers, and researchers to determine values). The ranking values have been set according to the highest number of classes in the indicator, where the low values indicate to low degradation or not degradation, while high values indicate to high degradation, fourteen parameters were used in ArcView GIS to produce the degradation-degree index. Where 34% of total study area was found had high degradation, 21% was low, while 45% was moderate degradation. it was observed that the deterioration is concentrated at the bottom of the study area and in the middle, and in addition to the loss in the percentage of vegetation as a result of degradation.

Database was built by collecting data and interviews the farmers through implementation of field work to reflect the nature and type of data. Samples were collected from the wells and analyzed to check the ionic characteristics of the elements and to estimate the proportion of sodicity as well as to measure the pH and EC values. On the other hand, soil samples were analyzed throughout 2012 and 2013 to monitor farming practices on cultivated land and the effects of salinization and fertilizers, also on abandoned and protected lands – and to assess soil quality through the analysis of its physical (its AH horizon, texture, structure, and color) and chemical characteristics (trace elements, pH and EC, and an estimate of the proportion of organic material). The results of these data show that human activities are most important where they have an impact on land degradation. Many factors impacted on the change in land use during the past twenty years where the barren soil area increased in 2010 to reach 143 km^2 after it had been about 88 km^2 in 1990, where the changes was validated by using remote sensing techniques.

The Second Objective

In **Chapter 5**, there was an attempt to map the groundwater vulnerability of the study area by using GIS and the DRASTIC model. Six DRASTIC parameters were used in ArcView GIS to produce the vulnerability index. It was found that 20.8% of the study area had high groundwater vulnerability while 31.4% had a low and about 47.8% moderate vulnerability. The land-use map of the study area was included in the DRASTIC index to investigate groundwater vulnerability. It was also found that 11.4% of the study area had a high vulnerability to possible sources of contamination, while less than 3% had a moderate vulnerability and a possible source of contamination by salinization and fertilizers.

The Third Objective

In **Chapter 5**, the quality of groundwater was predicted and evaluated through an analysis of water samples from the wells, where the spatial relationships between well locations and groundwater vulnerability showed most of the wells in the study area had high values of SAR and salinity.

The Fourth Objective

Due to the results that showed the existence of land deterioration as well as groundwater vulnerability to contamination in the study area, a relationship was appointed between the vulnerability of groundwater and land degradation due to hot-spot areas, where most of them are located in the middle and southeast of the study area, the total area of hot-spots being about 7.2 km^2, which reflects the key points that will help reduce pollution and degradation through the outlook proposed and accepted with consideration to size of problem.

References:

A

AARDO. 2007. Jordan Accession to WTO: Impact on Agriculture Sector. Afro-Asian Rural Development Organization (AARDO).New Delhi. No.10-07-300.

Abbas, I., and Fasona, M., 2012. Remote Sensing and Geographic Information Techniques: Veritable Tools for Land Degradation Assessment. American Journal of Geographic Information System. 1(1): 1-6. DOI: 10.5923/j.ajgis.20120101.0.

Aksoy, M. Atam., and Beghin, John, C., 2005. Global Agriculture Tural Trade and Developing Countries. The International Bank for Reconstruction and Development. The World Bank.

Al-Adamat, R, N, A., 2002. The use of Geographical Information Systems (GIS) and remote sensing to investigate Groundwater Quality in the Azraq Basin, Jordan. Unpublished PhD Thesis, Coventry University, Coventry UK.

Al-Ansari, N, A., and Baban, S, M, J, m., 2001. The climate and water resources. In Baban, S. M. J. and Al-Ansari, N. A. (eds.), Living with water scarcity: Water resources in Jordan, Badia Region, The way forward. Al al- Bayt University, Jordan.

Albaladejo, J., Martinez-Mena, M., Roldan, A and X, Castillo., 1998. Soil degradation and desertification induced by vegetation removal in a semiarid environment. Suil Use and Management. 14, l-5.

Al-Khaier, F., 2003. Soil Salinity Detection Using Satellite Remote Sensing, Master thesis. International Institute Geo-information Science and Earth Observation. Netherlands.

Al-Kilani, M, Mutasim., 2008. National Focal Point for Sustainable Development. CSD Guidelines For national Reporting to CSD-16. Department of Agricultural Policies and International Directorate. Jordan.
http://www.un.org/esa/agenda21/natlinfo/countr/jordan/.

Aller, L., Bennet, T., Lehr, JH., Petty, RI., and Hackett, G., 1987. DRASTIC: astandarized system for evaluating groundwater Agency. Report EPA/600/2- 87/036, p 643.

Allison, R, J. *et al.* 1998. Geology, Geomorphology, Hydrology, Groundwater and Physical Resources. In Dutton, R, Clarke, J and Battikhi, A, (eds.), Arid Land Resources and their Management, Jordan Desert Margin, Kegan Paul International, London, 332 p.

Al-Qudah, B., 2001. Soils of Jordan. In Zdruli S, Montanarealla L (eds), Soil Resources of Southern and Eastern Mediterranean Countries: CIHEAMIAMB, pp 127-141.

Al-Smadi, A., 1997. Geological Map of Al-Mafraq. Geological Mapping Division. National Mapping Project. Natural Resources Authority Geological Doctorate. 3254 IV. Scale 1:50. 000.

Altwegg, R., and Anderson, M, D., 2009., Rainfall in arid Zones: Possible effects of Climate Change on the Population Ecology of Blue Cranes. Functional Ecology. DOI: 10.1111/j.1365-2435.2009.01563.x.

Awawdeh, M., and Jaradat, R., 2010. Evaluation of aquifers vulnerability to contamination in the Yarmouk River basin, Jordan, based on DRASTIC method. Arab J Geosci. 3: 273–282.

B

Baatz, M. etal., 2004. eCognition Professional User Manual. Munich : Definiens Imaging GmbH.

Barrett, E., C., and Curtis, L, F., 1999. Introduction to environmental remote sensing, (4th edition). Stanley Thrones (Publishers) ltd, UK.

Barry, S., 1996. Monitoring Vegetation Cover. Conservationist. Alameda County Resource Conservation District. Holmes St., Livermore. CA 94550.

Bashur, E., and Antoine Al-Sayeg, A., 2007. Soil analysis methods arid areas and semi-arid. FAO. ISBN 978-92-5-6056611-0. American University-Beirut

Bentor, Y. K., 1986. The crustal evolution of the Arab-Nubian Massif with special reference to the Sinai Peninsula. Precam. Res. 28, 1-74.

Bergmann, W., 1992. Nutritional disorders of plants (development, visual and analytical diagnosis). Gustav Fischer Verlag Jena, Stuttgart, Germany. ISBN 3-334-60422-5.

Boer, MM., 1999. Assessment of dryland degradation: linking theory and practice through site water balance modeling. PhD, Utrecht University, The Netherlands, 288 pp.

Boyd, C.E., C.W. Wood, and T, Thunjai., 2002. Pond soil characteristics and dynamics of soil organic matter and nutrients. In: K. McElwee, K. Lewis, M. Nidiffer, and P. Buitrago (Editors), Nineteenth Annual Technical Report. Pond Dynamics/Aquaculture CRSP, Oregon State University, Corvallis, Oregon, pp. 1–10.

C

Campbell, J, B., 1996. Introduction to Remote Sensing (2nd Ed), London:Taylor and Francis.

Christensen, B,T., and Johnston, A, E., 1997. Soil organic matter and soil quality - lessons learned from long-term experiments at Askov and Rothamsted, in Gregorich, E.G. und Carter, M.R., (Hrsg.). Soil quality for crop production and ecosystem health, Seite(n) pp. 399-430. Developments in soil science 25. Elsevier Science.

Churchill, J., 2003. Change Detection of Mining and Urban Growth in South western West Virginia. Project for Advanced Remote Sensing (Geog/Geol 755). Dept. Geology and geography, West Virginia University.

Civco, D, L., Hurd, J, D., Wilson, E., Song, M., and Zhang, Z., 2002. A comparison of land use and land cover change detection methods. ASPRS-ACSM Annual Conference and FIG XXII Congress April 22-26.

Civita, M., 1994. Le carte della vulnerabilita` degli acquiferi all´inquinamiento. Studi sulla vulnerabilita` degli acquiferi (Map of aquifers' vulnerability to pollution. Aquifers' vulnerability study). Pitagora Editrice, Bolonia, p 326.

Clesceri, L, S., Greenberg, A, E., and Eaton, A, D., 1998. Standard Methods for the Examination of Water and Wastewater. 20th Edition.

Congalton, Russell G., 1991. A Review of Assessing the Accuracy of Classifications of Remotely Sensed Data. REMOTE SENS. ENVIRON. 37:35-46.

Contador, J, F., Schnable, S., Gutiérrez, A., and Fernández, M., 2009. Mapping Sensitivity to Land Degradation Extremadura. SW Spain. Land Degradation and Development. 20: 129-144.

Coppin, P., Jonckheere, I., Nackaerts, K., Muys, B., and Lambin E. 2004. Digital change detection methods in ecosystemmonitoring: A review, Int. J. Remote Sens. 25, 1565–1596.

Corwin, D, L., Lesch, S, M., Shouse, P, J., Soppe, R., and Ayars, J, E., 2003. Identifying soil properties that influence cotton yield using soil sampling directed by apparent soil Electrical conductivity. Agronomy Journal 95:352-364.

Coxon, C., and Thorn, R, H., 1989. Temporal variability of water quality and the implications for monitoring programmes in Irish limestone aquifers. In Saahuquillo, A., Anreu, J., and O'Donnell, T. (eds.), Groundwater Management Quantity and Quality. IAHS Publication No. (188), pp 111-120.

D

Da'as, A., and Walraevens, K., 2010. Groundwater Salinity in Jericho area, west Bank, Palestine. SWIM21-21st Salt Water Inrusion Meeting. Azores. Portugal.

Daniels, W. L., Zelazny, L,W., and Everett, C, J., 1987. Virgin hardwood forest soil of the southern Appalachian Mountains: II. Weathering, mineralogy, and chemical properties. Soil Sci. Soc. Am. J. 51, 730– 738.

Darwich, T., 2009. Sustainable land management practices to reverse land degradation in Lebanon. Economic and Social Commission for Western Asia (ESCWA). 09-0157.

De Chazal, J., and Rounsevell, M, D., 2009. Land-use and climate change within assessments of biodiversity change: A review. *Global Environmental Change, 19*(2), 306-315.

DESIRE, 2007. Disseminating Educational Science Innovation & Research in Europe.

http://www.desire-his.eu/wimba/WP2.1%20Indicators%20in%20the%20study%20sites%20(Report%2066%20D211%20Mar10)/index.htm

Dhembare, J, A., 2012. Assessment of Water Quality Indices for Irrigation of Dynaneshwar Dam Water, Ahmednagar, Maharashtra, India. Scholars Research Library. ISSN 0979-508X. Vol 4 No. (1). Pp 348-352.

Doerfliger, N., Jeannin, P-Y., and Zwahlen, F., 1999. Water vulnerability assessment in karst environments: a new method of defining protection areas using a multi- attribute approach and GIS tools (EPIK method).Environmental Geology;39(2):166-76.

Dong, L., Wang, W., Ma, M., Kong, J., and Veroustraete, F. 2009. The change of land cover and land use and its impact factors in upriver key regions of the Yellow River. *International Journal of Remote Sensing, 30*(5), 1251-1265.

DPP, 2007. Drynet Position Paper. Desertification and climate change: linkages, synergies and challenges. United Nation Convention to Combat Desertification.

Dunjó, G., Pardini, G., and Gispert, M. 2004. The role of land use-land cover on runoff generation and sediment yield at a microplot scale, in a small Mediterranean catchment. Journal of Arid Environments, 57(2), 239e256.

Duqqah, M., Naber, S., and Shatanawi M., 2007. Agriculture and irrigation water policies toward improved water conservation in Jordan. In : Karam F. (ed.), Karaa K. (ed.), Lamaddalen a N. (ed.), Bogliotti C. (ed.). Harmonization and integration of water saving options. Convention and promotion of water saving policies and guidelines. Bari : CIHEAM / EU DG Research. Page. 97 -109 (Option s Méditerran éen n es : Série B. Etu des et Rech erch es; n . 59).

Dur,. Dennis C., Franklin, Steven E., and Dube. Monique G., 2012. A comparison of pixel-based and object-based image analysis with selected machine learning algorithms for the classification of agricultural landscapes using SPOT-5 HRG imagery. Journal of remote sensing of environment 118. 259–272.

E

Essa, S., 2004. GIS Modelling of Land Degradation in Northern-Jordan Using Landsat Imagery. International congress for photogrammetry and remote sensing; ISPRS XXth congress. ISSN: 1682-1750.

Evans, BM., and Myers, WL., 1990. A GIS-based approach to evaluating regional groundwater pollution potential with DRASTIC. J Soil Water Conserved. 45:242– 5.

F

Fadhil, A., 2009. Land Degradation Detection Using Geo-Information Technology for Some Sites in Iraq. Journal of Al-Nahrain University Vol.12 (3). pp. 94-108.

FAO. 1979. Food and Agricultural Organization of the United Nations. A provisional methodology for soil degradation assessment. Rome.

FAO (Food and Agriculture Organization). 1987. Soil and Water Conservation in semi-arid areas. Prepared By Hudson, N, W., Ampthil, S, A., and Bedford United Kingdom. Producedby: Natural Resources Management and Environment Department:

http://www.fao.org/docrep/T0321E/T0321E00.htm

FAO. 1999. Drylands and the MFCAL approach in cultivating our futures – Background papers prepared for The FAO/Netherlands Conference on the Multifunctional Character of Agriculture and Land (MFCAL), 12-17 Sep 1999, Maastricht, The Netherlands. Rome, Italy: FAO.

FAO. 2006. Guideline for soil Description. ISBN 92-5-105521-1.

FAO. 2006. Irrigation and Drainage Paper: Water Quality for Agriculture. Guidelines for irrigation water quality. 29.1. (Water Wells and Borehole s B. Misstear, D. Banks and L. Clark. John Wiley & Sons, Ltd). ISBN: 0-470-84989-4.

http://onlinelibrary.wiley.com/doi/10.1002/0470031344.app3/pdf

FAO. 2009. High Level Expert Forum - How to Feed the World in 2050. Office of the Director, Agricultural Development Economics Division. Economic and Social Development Department. Viale delle Terme di Caracalla, Rome. Italy.

FAO. 2011. Mitigation of Climate Change in Agriculture Series 3. Climate-Smart Agriculture: A Synthesis of Empirical Evidence of Food Security and Mitigation Benefits from Improved Cropland Management:

http://www.fao.org/docrep/015/i2574e/i2574e00.pdf

FAO. 2012. The State of Food and Agriculture, Food and Agriculture Organization of the United Nation. ISSN 0081-4539.

Ferrara, G., Farrag, K., and Brunetti, G., 2012. The effects of rock fragmentation and / or deep tillage on soil skeletal material and chemical properties in a Mediterranean climate. Soil Use and Management. 28. 394–400.

Foster, S., 1987. Fundamental concepts in aquifer vulnerability, pollution risk and protection strategy. In: Van Duijvenbooden W, Van Waegeningh HG (eds) Vulnerability of soil and groundwater to pollution, vol 38. TNO committee on hydrological research, La Haya, pp 69–86.

G

GCEP (General Corporation for the Environment Protection). 2001. The General Corporation for the Environment Protection. Conservation and Sustainable Use of Biological Diversity in Jordan. First National Report of the Hashemite Kingdom of Jordan on the Implementation of Article 6 of the Convention on Biological Diversity.

Gessler, P, E., Moore, I, D., McKenzie, N, J., and P. J. Ryan 1995. Soillandscape modeling and spatial prediction of soil attributes, Int. J. Geogr. Inf. Syst., 9, 421- 432, doi:10.1080/02693799508902047.

Gilley, J, E., and Doran, J, W., 1997. Tillage Effects on Soil Erosion Potential and Soil Quality of a Former Conservation Reserve Program Site. Biological Systems Engineering: Papers and Publications. Paper 138.

Goldscheider, N., Klute, M., Sturm, S., and Ho¨ ltz, H., 2000. The PI method—a GISbased approach to mapping groundwater vulnerability with special consideration of karst aquifers. Z Angew Geol 46(3):167–166.

Grace, G., 1993. Start with the Soil. Soils and Soil Physical Properties. Pp. 27-38.

Gretton, P., and Salma, U., 1997. Land Degradation: Links to Agricultural Output and Profitability. Australian journal of Agricultural and Resource Economics, Vol. 42, No. 2, pp 209-225.

GTZ, and Agrar und Hydrotechnik (AHT Group). 1977. The national water master plan [Seven volumes]. Amman, Jordan.

H

Havlin, J, L., James, D, B., Samuel, L, T., and Werner, L, N., 1999. Soil fertility and fertilizers. An introduction to nutrient management. Prentice Hall upper Saddle River, New Jersey, Sixth Edition, ISBN (0-13-626806-4).

Heiniger, R, W., McBride, R, G., and Clay, D, E., 2003. Using soil electrical conductivity to improve nutrient management. Agronomy Journal 95:508-519.

Herrick, J, E., Tugel, A, J., Shaver, P, L., and Pellant, M., 2001. Rangeland soil quality: organic matter. Soil Quality Information Sheet. Rangeland Sheet 6.

Hjort, K., USAID., AMIR Project., Zakaria, M., and Falah I, Salah (MoA)., 1998. An Introduction to Jordan's Agriculture Sector and Agricultural Policies. WTO Accession Unit. Ministry of Industry and Trade. Jordan.

Hockensmith, RD., Steele, JG., 1949. Recent Trends in the Use of the Land-Capability Classification. Proceedings Soil Science Society of America, 14, 383-388.

Horneck, D, S., Ellsworth, J, W., Hopkins, B, G., Sullivan, D, M., and Stevens, R, G., 2007. Managing Salt-Affected Soils for Crop Production. PNW 601-E. Oregon State University, University of Idaho, Washington State University.

I

Imeson, AC., Emmer, I., 1992. Implications of climatic change for land degradation in the Mediterranean. In: Climatic change in the Mediterranean. Environmental and social impacts of climatic change and sealevel rise in the Mediterranean region. (eds Jeftic L, Milliman JD, Sestini G)., Edward Arnold, London.

Ibrahim, M., 2010. Environmental Assessment of Land Use Changes Using Remote Sensing and GIS Techniques in Irbid Area. Master thesis unpublished. Al al-Bayt University – Jordan.

IPCC. 2007. Climate Change 2007. Impacts, adaptions and vulnerability. In M. L. Parry, O. F. Canziani, J. P. Palutikof, P. J. van der Linden, & C. E. Hanson (Eds.), Contribution of working group II to the fourth assessment, report of the intergovernmental panel on climate change. Cambridge, UK: Cambridge University Press.

IPCC. 2008. Climate Change and Water, Intergovernmental Panel on Climate Change Technical Report IV. June 2008.

J

Javadi, S., Kavehkar, N., Mousavizadeh, M, H., and Mohammadi, K. 2011. Modification of DRASTIC Model to Map Groundwater Vulnerability to Pollution Using Nitrate Measurements in Agricultural Areas. J. Agr. Sci. Tech. Vol. 13: 239-249

Jankauskas, B., Jankauskienė, G., and Fullen M, A., 2007. Relationships between soil organic matter content and soil erosion severity in Albeluvisols of the Žemaičiai Uplands. EKOLOGIJA. Vol. 53. No. 1. P. 21–28

Jessica, E, L., and Sonia, T., 2009. Groundwater Vulnerability Assessments and Integrated Water Resource Management. Streamline Watershed Management Bulletin Vol. 13/No.

Jimoh, H, I., Ajewole, O, D., Onotu, S, I., and Ibrahim, R, O., 2011. Implications of Land Degradation, Reclamation and Utilisations in the Oil Producing Areas of Nigeria: Perspectives on Environmental Sustainability and Development. nternational Journal of Business and Social Science. Vol. 2 No. 22.

Jezierska, K., Gonet, B., Podraza, W., and Domek, H., 2011. Technical note A new method for the determination of water quality. Water SA Vol. 37 No. 1.

JISM (Jordanian Institute of Standards and Metrology). 2001 Technical Regulation – Water – Drinking Water: JS286, Amman, Jordan.

Jones, J, B., 1998. Plant nutrition manual. CRC Press LLcC, USA. ISBN 1- 884015-31-x.

Jong, Rogier de., Bruin, Sytze de., Schaepman, Michael E., Dent, David L., 2011. Quantitative mapping of global land degradation using Earth observations. International Journal of Remote Sensing, 32. pp. 6823-6853. DOI: 10.1080/01431161.2010.512946.

K

Kassas, M., 1995. Desertification: a general review. Journal of Arid Environments, 30, 115-128.

Klingebiel, AA., and Montgomery, PH., 1961. Land- Capability Classification, Handbook 210. In: US Dept Agriculture. Washington DC.

Kirk, A., 1998. The Effect of Intensive Irrigated Agricultural upon Soil Degradation: A case study from Ashrafiyya. In Dutton, R, Clarke, J and Battikhi, A, (eds.), Arid Land Resources and their Management, Jordan Desert Margin. Kegan Paul International, London.

Kitchen, N, R., Drummond, S, T., Lund, E, D., Sudduth, K, A., and Buchleiter, G, W., 2003. Soil Electrical conductivity and topography related to yield for three contrasting soil-crop systems. Agronomy Journal 95:483-495.

Knox, R. C., Sabatini, D. A., and Canter, L. W., 1993. Subsurface transport and fate processes. USA: Lewis Publishers.

Kosmas. C., Kirkby. M., and Geeson, N., 1999. The Medalus project Mediterranean desertification and land use. Manual on key indicators of desertification and mapping environmentally sensitive areas to desertification. European Commission. EUR 18882.

Khresat, S. A. Rawajfih, Z. and Mohammad, M. 1998. Land degradation in north-western Jordan: causes and processes. Journal of Arid Environments. 39: 623–629.

Kumar, S., Thirumalaivasan, D., Radhakrishnan, N., 2013. GIS Based Assessment of Groundwater Vulnerability Using Drastic Model. Arabian Journal for Science and Engineering, Volume 39, Issue 1, pp 207-216.

Kundzewicz, Z., W., Mata, L, J., Arnell, N. W. et al., 2007. Freshwater resources and their management. In Climate Change: Impacts, Adaptation and Vulnerability. Contribution of Working Group II to the Fourth Assessment Report of the Intergovernmental Panel on Climate Change, ed. M. L. Parry, O. F. Canziani, J. P. Palutikof, P. J. van der Linden and C. E. Hanson, 173–210. Cambridge University Press.

L

Lal R., Kimble, J, M., Follett, R, F., and Stewart, B, A., 1998. Soil Processes and the Carbon Cycle. Boca Raton, Florida: CRC Press. 609 p.

Lee, J, Y., and Warner, T, A., 2004. Image classification with a regional based approach in high spatial resolution imagery. XXth ISPRS Congress, Istanbul, Turkey.

http://www.isprs.org/istanbul2004/comm3/papers/438.pdf

Lillesand, T, M., and Kiefer, R, W., (2001) Remote Sensing and Image Interpretation, 4th ed, John Wiley and Sons, inc. USA, ISBN: 0471255157.

Lindenmayer, D., Burgman, M., 2005. Practical Conservation Biology. Vegetation Loss and Degradation. Chapter 9. P. 229-254.

Li, X, Y., Ma. Y, J., Xu, H, Y., Wang, J, H., and Zhang, D, S., 2009. Impact of land use and land cover change on environmental degradation in Lake Qinghai watershed, northeast Qinghai-Tibet Plateau. *Land Degradation & Development, 20*(1), 69-83.

Logsdon, S, D., Prueger, H, J., Meek, W, D., Colvin, S, T., Milner, M., and James, E, D., 1998. Crop yield variability as influenced by water in rain-field agriculture. p. 453-465. In P. C. Robert et al. (ed.) Proc. 4 th Int. Conf. on Precision Agriculture. American Society of Agronomy, Madison, WI.

Louwagie, Geertrui., Gay, Stephan, H., Burrell, Alison., 2009. SoCo team. Report on the project 'Sustainable Agriculture and Soil Conservation (SoCo)' Addressing soil degradation in EU agriculture: relevant processes, practices and policies. Agriculture and Rural Development. European Commission. ISBN 978-92-79-11358-1.

Lu. D, P. Mausel, Brondízios. E, Moran. E., 2004. Change detection techniques, Int. J. Remote Sens. 25, 2365– 2407.

Lunetta, R, S., Elvidge, C, D., 1998. Remote Sensing Change Detection: Environmental Monitoring Methods and Applications (Ann Arbor Press, Chelsea).

M

Makhlouf, I., Abu-Azzam, H., and Al-Hiayri, A., 1996. Surface and subsurface lithostratigraphic relationships of the Cretaceous Ajlun Group in Jordan. Natural Resources Authority, Subsurface Geology Division, Bulletin 8.

Malambo, L., 2009. A Region based Approach to Image Classification, Proceedings Applied Geo-informatics for Society and Environment (AGSE), Stuttgart Germany, 13-18 July 2009. Mas, J.-F., 1999.

Monitoring land-cover changes. A comparison of change detection techniques, Int. J. Remote Sens. 20, 139–152 .

McCalla, Allex, F., 2001. Challenges to World Agriculture in the 21st Century. Agriculture and Resource Economics. University of California. Davis. Vol 4. No.

MEA (Millenium Ecosystem Assessment). 2005. Ecosystems and human well-being: synthesis. In: Millennium Ecosystem Assessment. Washington, DC., World Resources Institute / Island Press.

Maliva, R., and Missimer, T., 2012. Arid Lands Evaluation and Management. Environmental Science and Engineering. DOI: 10.1007/978-3-642-29104-3_2.

Merchant, J, M., 1994. GIS-Based groundwater pollution hazard assessment: A critical review of the DRASTIC Model. Photograommetric Engineering & Remote sensing, Vol. (60), No. (9), pp. 1117-1127.

Mhangara, Paidamwoyo., 2011. Land Use/Cover Change Modelling and Land Degradation Assessment in the Keiskamma Catchment Using Remote Sensing and GIS. PhD-thesis-Nelson Mandela Metropolitan University.

Mills, H, A., and Jones, J, B., 1996. Plant analysis handbook II (A practical sampling, preparation, analysis, and interpretation guide). MicroMacro Publishing, Inc, USA. ISBN 1-878148-052.

Mingguo, Z., Qiangguo, C., and Hao, C., 2007. Effect of vegetation on runoff-sediment yield relationship at different spatial scales in hilly areas of the Loess Plateau, North China. Acta Ecologica Sinic., 27(9). 3572–3581.

MoA (Ministry of Agriculture). 1993. National soil map and land use project. The soils of Jordan. Hunting Technical services Ltd. In association with Soil Survey and Land. Research Center. Vol. 2 Level 1.

MoA (Ministry of Agriculture). 2011, Annual Report.

Moh'd, B., 2000. The geology of Irbid and Ash Shuna Ash Shamaliyya (Waqqas). Map Sheet No. 3154-II and 3154-III. Natural Resources Authority, Geological Mapping Division, Bulletin 46.

Mouat, D, A., Mahin, G, G., Lancaster, J., 1993. Remote sensing techniques in the analysis of change detection, Geocarto Int. 2, 39–49.

MPRA (Munich Personal RePEc Archive). 2006. MCS, Bantilan., P, Anand Babu., GV, Anupama., H, Deepthi., and R, Padmaja., Dryland Agriculture: Dynamics, Challenges and Priorities. International Crops Research Institute for the Semi-Arid Tropics (ICRISAT). MPRA Paper No. 16423.

Munshower, F, F., 1994. Practical Handbook of Disturbed Land Revegetation. Lewis Publisher, Boca Raton, FL.

Muriuki, G., Seabrook, L., McAlpine, C., Jacobson, C., Price, B., and Baxter, G., 2011. Land cover change under unplanned human settlements: A study of the Chyulu Hills squatters, Kenya. *Landscape and Urban Planning, 99*(2), 154-165.

Murray, K, S., and Rogers, D, T., 1999. Groundwater vulnerability, Brownfield Redevelopment and Land Use Planning. Journal of Environmental Planning and Management. Vol. (42), No. (6), pp. 801-810.

Muthanna, N, M., 2011. Quality Assessment of Tigris River by using Water Quality Index for Irrigation. European Journal of Scientific Research. ISSN 1450-216X Vol.57 No. (1). Pp.15-28.

Muttitanon, W., and Tripathi, N, K., 2005. Land use/land cover changes in the coastal zone of Ban Don Bay, Thailand using Landsat 5 TM data. International Journal of Remote Sensing, Volume 26, Number 11, June 2005, pp. 2311-2323(13).

MWI (Ministry of Water and Irrigation). 2002. Water resources in Jordan. Found at: http://www.mwi.gov.jo/ministry of water and irrigationout.htm.

MWI (Ministry of Water and Irrigation). 2004. Annual report. Amman, Jordan.

MWI (Ministry of Water and Irrigation). 2011. Annual water budget report. Amman, Jordan.

N

Navarrete, C, M., Olmedo, J, G., Valsero, J, J, D., Go´mez, J, D, G., Espinar, J, A, L., and Go´mez, J, A, O., 2008. Groundwater protection in Mediterranean countries after the European water framework directive. Environ Geol. 64:637–649.

Navulur, Kumar., 2007. Multispectral Image Analysis Using Object Oriented Paradigm. New York : Taylor & Francis Group.

Ngamabou, René Siwe., 2006. Evaluating the Efficacy of Remote Sensing Techniques in Monitoring Forest Cover and Forest Cover Change in the Mount Cameroon Region. PhD thesis. Freiburg University. Germany.

Nortcliff, S., Carr, G., Potter, R, B., and Darmame. K., 2008. Jordan's Water Resources: Challenges for the Future. Geographical Paper No. 185. University of Reading. United Kingdom.

NRCS (Natural Recourses Conservation Service). 2011. NRCS East National Technology Support Center, NRCS National Soil Survey Center, ARS National Laboratory for Agriculture and the Environment, NCERA-59 Scientists, and Department of Natural Resources and Environmental Sciences, University of Illinois at Urbana-Champaign: http://soilquality.org/practices/tillage.html

NRCS (Natural Resources Conservation Service). 2012. USDA (United States Department of Agriculture). Soil qulity, Soil pH. Through Last Modified: 10/16/2012. URL: http://soils.usda.gov/sqi/assessment/educators.html.

O

Oppenheim, A, N., 1992. Questionnaire design, interviewing and attitude measurement, (New edition). Pinter Publishers, London.

P

Parker, D., 1970. The hydrogeology of the Mesozoic-Cenozoic aquifers of the western highlands and plateau of east Jordan. Investigation of the sandstone aquifers of east Jordan, Technical Report No. 2: UNDP/FAO Project 212.

Peterson, D, L., Egbert, S.L., Price, K.P. and Martinko, E. A., 2004. Identifying historical and recent land-cover changes in Kansas using postclassification change detection techniques. Transactions of the Kansas Academy of Science: Vol. 107, No. 3, pp. 105–118.

Piscopo, G., 2001. Groundwater vulnerability map, explanatory notes, Castlereagh Catchment, NSW Department of Land and Water Conservation, Australia.

Q

R

Raddad, K., 2005. Water supply and water use statistics in Jordan. IWG-Env, International Work Session on Water Statistics, Vienna.

Renard, K.G., Foster, G.R., Weesies, G.A., McCool, D.K., and Yoder D.C. (1997). "Predicting soil erosion by water: A guide to conservation planning with the Revised Universal Soil Loss Equation (RUSLE)." Agriculture Handbook No.

Richards, John A., and Jia. Xiuping., 2006. Remote Sensing Digital Image Analysis an Introduction. 4th Edition. ISBN-13 978-3-540-25128-6 Springer Berlin Heidelberg New York.

Rosen, L, A., 1994. Study of the DRASTIC methodology with emphasis on Swedish conditions. Ground Water; 32(2):278 –85.

Rusu, T., Moraru, P, I., Bogdan, I., Pop, A, I., and Sopterean, M, L., 2011. Effect of Soil Tillage System upon the Soil Properties, Weed Control, Quality and Quantity Yield in Some Arable Crops. World Academy of Science. Engineering and Technology 59.

S

Salameh, E., Al-Ansari, N., and Al-Nsoor, I., 1997. Water and Environment in the Area East of Mafraq and their Developmental Potentials. Report No. (1). Strategic Environment and Water Resources Unit, AL Al-Bayt University, Jordan.

Salameh, E., Bannayan, H., 1993. Water Resources of Jordan, Present status and Future Potentials. Friedrich Ebert Stiftung, Amman, 183 p.

Samuel, A-B., 2010. Land Management Practices and Their Effects on Food Crop Yields in Ghana. 3rd African Association of Agricultural Economists (AAAE) and 48th Agricultural Economists Association of South Africa (AEASA) Conference.

Scheer. S. J., 1999. Soil Degradation a Threat to Developing-Country Food Security by 2020. Food, Agricultural, and the Environment Discussion Paper 27. International Food Policy Research Institute. USA.

Seelig, B, D., 2000. Salinity and Sodicity in North Dakota Soils. EB-57. North Dakota State University, Fargo, ND.

Senjobi, B, A., and Ogunkunle, A, O., 2011. Effect of different land use types and their implications on land degradation and productivity in Ogun State, Nigeria. Journal of Agricultural Biotechnology and Sustainable Development Vol. 3(1) pp. 7-18.

SFSA (Syngenta Foundation for Sustainable Agriculture) 2013. WRO-1002.11.51 Postfach. CH-4002 Basel. Switzerland. http://www.syngentafoundation.org/index.cfm?pageID=45.

Shatanawi, M., Al-Weshah, R., and Al-Ayed, R., 1999. Relationship between Surface and Groundwater for Artificial Recharge, Internal Report, Jordan Badia Research and Development Programme, Amman, Jordan.

Shiklomanov, I.A., and Rodda, J, C, Eds., 2003. World Water Resources at the Beginning of the 21st Century. Cambridge University Press, Cambridge, 435 pp.

Shirazi, S, M., Imran, H, M., Akib, Shatirah., 2012. GIS-based DRASTIC method forgroundwater vulnerability assessment:a review. Journal of Risk Research Vol. 15, No. 8.

Singh, A., 1989. Digital change detection techniques using remotely-sensed data, Int. J. Remote Sens. 10, 989– 1003.

Sivakumar, M, V, K., 2007. Interactions between climate and desertification. Agricultural and Forest Meteorology 142. 143–155.

Smaeil, Z. H., 2002. Marl Soil and different types of erosion in Iran. Symposium no. 37. Paper no. 2235.

Sokouti, R., Mahdian, M, H., and Farshad, A., 2009. The Effects of Physical and Chemical properties of marl derived soils on the erosion forms and rate. Goldschmidt Conference. A1246.

Solaiman, K., Modallaldoust, S., Lotfi, S., 2009. Investigation of land use changes on soil erosion process using geographical information system. Int. J. Environ. Sci. Tech., 6 (3), 415-424,

Sonon, Leticia, S., Saha, Uttam and Kissel, David, E., 2012. Soil Salinity Testing, Data Interpretation and Recommendations. Agricultural and Environmental Services Laboratories. College of Agricultural and Environmental Sciences. University of Georgia. Circular 1019.

Story, M., and Congalton, R, G., 1986. Accuracy assessment: A users perspective. Photogrammetric Engineering and Remote Sensing 52(3):397-399.

Stuyfzand., P, J., 1986. A new hydrochemical classification of watertypes: principles and application to the coastal dunes aquifer system of the Netherlands. Proceedings of the 9th Salt Water Intrusion Meeting, Delft, 641-655.

Summerfield, M, A., 1997. Global Geomorphology, 537 pp. Longman, New York.

SWP (Soil Web Project). 2004. Soil Web Project. Faculty of Land and Food Systems. The University of British Columbia.

http://www.landfood.ubc.ca/soil200/classification/soilformation_factors.htm#14

T

Ta"any, R., Batayneh, A., and Jaradat. R., 2007. Evaluation of Groundwater Quality in the Yarmouk basin Nourth Jordan, Journal of Environmental Hydrology,Volume 15.

Tardie. P. S. and Congalton. R. G. (2002): A Change Detection Analysis: Using Remotely Sensed Data to Assess the Progression of Development of the Essex County, Massachusetts from 1990 to 2001.

http://www.unh.edu/natural-resources/pdf/tardie-paper1.pdf.

Tesfa, T., Tarboton, D., Chandler, D., and McNamara, J., 2009. Modeling Soil Depth from Topography and Land Cover attributes. Water Resources Research. Vol. 45. W10438.

Théau, Jérôme., 2012. Springer Handbook of Geographic Information. Change Detection ISBN: 978-3-540-72678-4. pp 75-94.

TNAU (TamilNadu Agricultural University). 2008. TNAU Agritech Portal- Coimbatore. Agriculture-Tallige Types:

http://agritech.tnau.ac.in/agriculture/agri_tillage_typesoftillage.html

Toor, Gurpal, S., and Shober, Amy, L., 2008. Soils & Fertilizers for Master Gardeners: Soil Organic Matter and Organic Amendments. Institute of Food and Agricultural Sciences, University of Florida.

Tripepi, R, R., Bell, S., Bauer, M., and Jones, W., 2011. Master Gardener Program Handbook. Soil and fertilizer. University of Idaho.

Truscott, L., 2012. The Organic advantage through the lenses of the Eco-Index – Land Use Intensity:

http://info.textileexchange.org/te-farm-blog/bid/149806/The-Organic-advantage-through-the-lenses-of-the-Eco-Index-Land-Use-Intensity

Tsui, C-C., Chen, Z-S., Hsieh, C-F., 2004. Relationships between soil properties and slops position in a lowland rain forest od southern Taiwan. Geoderma 123. P 131-142.

U

UNCED (United Nations Conference on Environment and Development). (1992). "United Nations Conference on Environment and Development -Earth Summit", Rio de Janeiro, Brazil,

UNEP (United Nations Environment Programme). 1997. World atlas of desertification 2ED. UNEP, London.

UNEP (United Nations Environment Programme). 2007. GE04 Global Environment Outlook (environment for development), United Nations Environment Programme, New York.

UNCCD. 2004. "Fact sheets on UNCCD", URL: http://www.unccd.int/publicinfo/factsheets/menu.php

USAID. 2012. Review of Water Policies in Jordan and Recommendations Strategic Priorities. Final Report. Jordan Water Sector Assessment Program. ON. 01/AID-OAA-TO-10-00025. CN. #EPP-I-00-04-00023-00.

USDA (United State Department of Agriculture). 1987. Soil Mechanics Level I – Module 3. Soil Conservation Service. National Employee Development Staff.

USDA (United State Department of Agriculture). 1994. Permeability Key.

USDA (United State Department of Agriculture). 2001. USDA, Natural Resources Conservation Service. Rangeland Soil Quality—Organic Matter. Soil Quality Information Sheet. Rangeland Sheet 6.

USEPA (United State Environmental Protection Agencies). 1974. Quality criteria for water Ed. R. C. Trtain, Casste House, Publ. Great Britan.

UN Water. 2007. Coping with water scarcity: challenges of the twenty first century. Prepared for World Water Day, 2007.

http://www.unwater.org/wwd07/download/documents/escarcity.pdf

V

Vargas, RR., Omuto, CT., and Njeru, L., 2007. Land degradation assessment of a Selected Study Area in Somaliland: The application of LADAWOCAT approach. In: SWALIM L-10 Land Degradation Report. Nairobi, Kenya, Food and Agricultural Organization of the United Nations.

Vásquez, R., Ramos, E., Oleschko, K., Sandoval, L., Parrot, J., and Nearing, M, A., 2010. Soil erosion and runoff in different vegetation patches from semiarid Central Mexico. Catena 80. 162–169.

Verburg, P, H., Van de Steeg, J., Veldkamp, A., and Willemen, L., 2009. From land cover change to land function dynamics: A major challenge to improve land characterization. *Journal of Environmental Management 90*, 1327-1335.

Vı´as, J, M., Andreo, B., Perles, M, J., Carrasco, F., Vadillo, I., and Jime´nez, P., 2006. Proposed method for groundwater vulnerability mapping in carbonate (karstic) aquifers: the COP method. Application in two pilot sites in Southern Spain. Hydrogeol J doi:10.1007/ s10040-006-0023-6.

Vrba, J., and Zoporozec, A., [eds.]. 1994. Guide book on Mapping Groundwater Vulnerability. - International Contributions to Hydrogeology (IAH), 16: 131 p.; Hannover.

W

WAJ (Water Authority of Jordan). 2005. Internal files for groundwater basins in Jordan.

Walkley, A., 1947. A critical examination of a rapid method for determining organic carbon in soils—effect of variations in digestion conditions and of inorganic soil constituents. Soil Sci. 63:251-264.

Wang, H, Q., E, C, Ellis., 2005. Image misregistration error in change measurements, Photogram. Eng. Remote Sens. 71, 1037–1044.

WB (World Bank). 2006. Sustainable Land Management: Challenges, Opportunities, and Trade-offs. Washington, DC 20433. The International Bank for Reconstruction and Development/The World Bank.

WB (World Bank). 2008. Agriculture for Development. World development report. The International Bank for Reconstruction and Development. Washington, DC 20433.

WB (World Bank). 2011. The world Bank-Jordan. The data world bank.

http://data.worldbank.org/country/jordan

Wheater, S., Howard., Mathias, A., Simon, Li, X., 2010. Groundwater modelling in arid and semi-arid areas. Cambridge University Press. 978-0-521-11129-4.

White, R., Tunstall, D., and Henninger, N., 2002. An ecosystem approach to drylands: Building support for new development policies. Information Policy Brief no. 1. Washington DC, USA: World Resources Institute.

Wilkie, D, S., and Finn, J, T., 1996. Remote sensing for natural resources monitoring: a guide to first time users. Columbia University Press, New York.

WMO (World Meteorological Organization). 2006. "World meteorological organization-World day to combat desertification, Climate and land degradation". WMO-No. 989. Geneva. Switzerland, pp. 32.

X

Xian-Li, X., Ke-Ming, M., Bo-Jie, F., Xain-Chum, L., Yong, H., Jian, Q., 2006. Research Review of the Relationship between Vegetation and Soil Loss. Acta Ecolohica Sinica. Vol. 26. No. 9.3137-3143.

Y

Yu, Q., 2006. Object-based Detailed Vegetation Classification with Airborne High Spatial Resolution Remote Sensing Imagery. Photogrammetry Engineering & Remote Sensing, Vol. 72, pp. 799-811.

Z

Zanen, M., and Koopmans, C., 2005. Nitrogen efficiency in organic farming using a GPS precision farming technique. Researching Sustainable Systems - International Scientific Conference on Organic Agriculture, Adelaide, Australia, September 21-23.

Zhao, G., and Meng, Y., 2010. Remote Sensing Image Based Information Extraction for Land Salinized Degradation and Its Evolution. Sixth International Conference on Natural Computation. 978-1-4244-5961-2/10/$26.00 © IEEE.

Zomer, R, J., Trabucco, A., Bossio, D, A., Verchot, L, V., 2008. Climate change mitigation: A spatial analysis of global land suitability for clean development mechanism afforestation and reforestation. Agriculture, Ecosystems & Environment. Volume 126. Issues 1–2. Pages 67–80.

Appendix (A-I)

Research Questionnaire

Type 1: Farm and Well data

1. Farm

- Number of Farm

- Farm code (1) - Farm code (2)

- Start in agriculture date(s)

Farm (1)	/ /
Farm (2)	/ /

- Date for start getting the crops from the lands

Farm (1)	/ /
Farm (2)	/ /

- Times of irrigation Rate and amounts of water.

Farm (1)	
Farm (2)	

- Times Ventilation

Farm (1)	
Farm (2)	

- GPS Coordinates

	N			E		
	Degree	Minute	Second	Degree	Minute	Second
Farm (1)						
Farm (2)						

1. Well

- Number of wells that supply the farm with water

- Well code (1) - Well code (2)

Well (s) digging date(s)

Well (1)	/ /
Well (2)	/ /

Date for start cultivating the land using water from the well (s)

Well (1)	/ /
Well (2)	/ /

What is the depth to water surface?

Well (1)	m
Farm (2)	m

- GPS Coordinates

	N			E		
	Degree	Minute	Second	Degree	Minute	Second
Well (1)						
Well (2)						

Type 2: Farm data

A. Crop type and area

1. Trees Farm

Crop	Tree (number)

2. Vegetable farms

Crop	Area (Ha)

B. Fertiliser types and amounts

1. Organic and Inorganic fertilisers

Type (Commercial name)	Amount (kg ha-1)

2. Pesticide applications

Type (Commercial name)	Amount (kg ha-1)/ (Litre ha-1)

Type 3: The agricultural practices

Q1: Do you have any kind of cooperation with other farmers in the area regarding the timing of spraying the crops?

Q2: How many times do you change kind of crops? And how many times do you change place of farm?

Q3 if you change crops. Why are you changed kinds of crops and what are factors that depend in choice place of agriculture?

Q4: What are the reasons behind shifting the location of cropping annually?

(1) --
(2) --
(3) --
(4) --

Q5: Who provides you with the extension services regarding the use of fertilizers and pesticides?

Q6: What are the consequences of spraying your field with pesticides and your neighbor farmers do not do so?

Q7: How is the design of the farm, whether it is an important consideration for farm?

Q8: General notes about the farm and the well (s)

Type 4: Soil Questioner

No. Sample:... Name of sample (Code): ....

Location: .... Deep of Sample: ....

Coordinate system of Sample.

	N			E		
No. Sample	Degree	Minute	second	Degree	Minute	second

Method of collect sample: ....

The propose of collection sample: ....

Land use of area: ....

Years of using:..

Yield for this year: .... / Yield for last year: ....

Irrigation way: ....

Years of irrigation ....

Structure of sample: ....

Horizon of sample:..

Appendix (A-II)

Secondary Data

(A-II).1.Wells Data (Jordan Ministry of Water and Irrigation)

Station-ID	Palestine-N	Palestine-E	Well Depth	Test Data
AD1120	1205350	264940	240	05-Okt-67
AD1121	1207265	266440	248	28-Jun-71
AD1123	1207715	266340	215	21-Jun-75
AD1124	1207685	265980		25-Jun-76
AD1127	1207705	267553	377	01-Okt-69
AD1129	1212510	265300	313	03-Okt-76
AD1130	1212320	266200	300	31-Jan-78
AD1148	1203900	264400	250	18-Sep-71
AD1149	1208590	264890	289	09-Feb-70
AD1150	1206090	264865	265	31-Jan-70
AD1257	1208310	257355	350	29-Apr-79
AD1257	1208310	257355	350	30-Jun-08
AD1262	1211673	258976	385	11-Okt-82
AD1278	1206787	266652	443	21-Mai-91
AD1278	1206787	266652	443	01-Sep-10
AD1320	1212097	258752	357	05-Okt-91
AD1327	1212950	262550	264	01-Nov-87
AD3004	1214214	261936	275	13-Mrz-93
AD3005	1203100	274500	451	21-Dez-91
AD3005	1203100	274500	451	21-Dez-91
AD3027	1201351	264765	204	-
AD3030	1205223	271638	209	-
AD3031	1212874	262091	245	-
AD3040	1212097	258752	590	21-Jul-98
AD3040	1212097	258752	590	13-Jan-05
AD3056	1198707	271648	261	28-Apr-00
AD3057	1206555	265850	310	08-Apr-00
AD3057	1206555	265850	310	20-Mrz-10
AD3057	1206555	265850	310	08-Apr-00
AD3057	1206555	265850	310	20-Mrz-10
AD3061	1212175	259350	447	16-Jul-00

A-II: Continue

AD3063	1207690	265985	270	08-Aug-00
AD3067	1207283	264435	258	19-Jun-02
AD3077	1211644	262320	422	02-Jan-06
AD3078	1207990	266390	379	02-Jan-06
AD3082	1195645	264555	300	03-Mrz-05
AD3089	1198600	267900	400	13-Jun-07
AD3118	1212627	261552	422	28-Apr-10
AD3124	1207267	266415	404	08-Aug-10
AD3131	1210623	255426	490	09-Jan-11
AD3132	1211530	262236	480	09-Jan-11
AD3140	1207685	265988	423	01-Mrz-11
AL1012	1173575	272442	91	-
AL1022	1175329	273377	100	16-Jul-65
AL1023	1175165	272236	100	07-Aug-65
AL1026	1174611	273651	100	18-Dez-66
AL1031	1173884	274269	150	25-Sep-66
AL1032	1173938	274316	100	09-Feb-91
AL1036	1175008	273076	100	06-Mrz-72
AL1040	1177834	271278	165	19-Dez-66
AL1118	1174900	258403	116	09-Jun-81
AL1489	1191532	274643	290	20-Jan-90
AL1522	1192450	272800		07-Mai-98
AL1522	1192450	272800		07-Mai-98
AL2414	1178400	254670	426	09-Okt-89
AL2414	1178400	254670	426	09-Okt-89
AL2710	1191502	273462	400	-
AL3003	1190042	273118	386	29-Jan-92
AL3003	1190042	273118	386	29-Jan-92
AL3374	1195200	266600	354	11-Apr-93
AL3374	1195200	266600	354	11-Apr-93
AL3375	1188641	273193	382	19-Dez-95
AL3376	1187626	273393	342	25-Mrz-96
AL3377	1185931	274075	400	04-Apr-96

A-II: Continue

AL3377	1185931	274075	400	04-Apr-96
AL3394	1184790	265652	181	-
AL3397	1195045	274587	264	-
AL3463	1190295	273544	358	02-Dez-95
AL3467	1195200	266600	340	02-Mrz-93
AL3468	1192600	270400	412	31-Dez-95
AL3483	1179750	258400	292	17-Apr-98
AL3484	1180950	258350	300	17-Apr-98
AL3485	1180250	258950	300	22-Jun-96
AL3488	1175335	273377	150	11-Jan-97
AL3516	1192570	272735	330	22-Sep-98
AL3522	1185720	270500	490	
AL3626	1175350	271350	300	11-Sep-01
AL3647	1177285	263700	234	12-Jun-03
AL3657	1191510	273469	415	-
AL3682	1191510	273465	415	-
AL3705	1182130	271775	400	02-Jun-04
AL3707	1174502	273550	250	21-Mai-04
AL3713	1180717	256600	340	30-Jul-04
AL3713	1180717	256600	340	30-Jul-04
AL3791	1180847	259690	300	12-Apr-07
AL3839	1173415	269710	267	01-Mrz-08
AL3982	1189134	274900	450	13-Jun-11
AL4080	1190295	273554	434	08-Mrz-13

Appendix (B)

Primary Data Soil Data

Appendix (B-II). Soil Data

A. Electrical Conductivity – 2013

Sample Code	Analysis	Sample Code	Analysis	Sample Code	Analysis
Protected	EC ms/cm	Caltivated	EC ms/cm	Abandoned	EC ms/cm
H36A	10,31	H 11 A	4,98	H 2 A	13,44
H 19 A	12,8	H 15 A	0,596	H 12 A	0,495
H 1 A	5,27	H 17 A	50,2	H 16 A	4,3
H 60 A	0,233	H 31 A	0,792	H 20 A	0,755
H 59 A	1,411	H 33 A	0,601	H 23 A	7,18
H 21 A	11,7	H 34 A	0,471	H 25 A	0,408
H 13 A	0,345	H 35 A	1,557	H 26 A	3,95
H 47 A	1,379	H 37 A	0,284	H 27 A	0,899
H 56 A	0,406	H 39 A	4,73	H 29 A	1,281
H 18 A	5,28	H 41 A	0,693	H 32 A	0,872
H 42 A	17,22	H 45 A	2,38	H 40 A	14,34
H 14 A	14,94	H 49 A	0,316	H 43 A	19,07
H 30 A	0,366	H 50 A	0,498	H 46 A	15,45
H 55 A	1,647	H 52 A	0,38	H 48 A	2,14
H 44 A	5,61	H 53 A	0,583	H 54 A	1,252
H 77 A	0,233	H 57 A	0,545	H 66 A	1,407
H 28 A	1,476	H 61 A	0,256	H 67 A	0,513
H 64 A	18,21	H 62 A	0,458	H 68 A	1,998
H 65 A	0,753	H 63 A	9,37	H 70 A	2,14
H 38 A	2,2	H 69 A	2,38	H 71 A	19,07
H 58 A	0,428	H 74 A	0,601	H 72 A	0,428
H 76 A	18,21	H 75 A	0,545	H 73 A	1,411

Continue - Electrical Conductivity – 2012

Sample Code	Analysis	Sample Code	Analysis
Caltivated	EC ms/cm	Abandoned	EC ms/cm
H34 A	0,41	H 25 A	0,4
H 74 A	0,641	H 72 A	0,382
H 15 A	0,68	H 12 A	0,534
H 61 A	0,25	H 67 A	0,463
H 31 A	0,86	H 20 A	0,849
H 41 A	0,628	H 32 A	0,92
H 37 A	0,24	H 27 A	0,972
H 53 A	0,54	H 54 A	1,65
H 39 A	3,98	H 29 A	1,08
H 57 A	0,496	H 66 A	1,1
H 75 A	0,453	H 73 A	1,68
H 62 A	0,49	H 68 A	2,8
H 52 A	0,32	H 48 A	1,56
H 63 A	10,2	H 70 A	2,14
H 35 A	1,36	H 26 A	4,68
H 17 A	44,6	H 16 A	5,65
H 33 A	0,56	H 23 A	6,71
H 11 A	5,6	H 2 A	11,94
H 45 A	2	H 40 A	15,9
H 50 A	0,53	H 46 A	15,65
H 49 A	0,29	H 43 A	17,3
H 69 A	2,79	H 71 A	17,4

B. Organic Matter – 2013

Sample Code	Analysis	Sample Code	Analysis	Sample Code	Analysis
Protected	OM	Abandoned	OM	Caltivated	OM
H14 A	1,5	H16 A	1,08	H 17 A	0,51
H28 A	0,77	H 26 A	1,2	H 35 A	0,54
H 44 A	0,85	H 43 A	0,9	H 49 A	0,59
H 38 A	0,79	H 32 A	1,2	H 41 A	0,61
H 13 A	0,9	H 12 A	0,41	H 15 A	0,69
H 21 A	1,13	H 25 A	1,3	H 34 A	0,74
H 59 A	0,87	H 67 A	1,3	H 61 A	0,74
H 47 A	1,7	H 46 A	0,82	H 50 A	0,77
H 1 A	1,13	H 2 A	1,1	H 11 A	1
H 65 A	1,9	H 71 A	0,9	H 69 A	1,05
H 42 A	1,13	H 40 A	1,2	H 45 A	1,05
H 56 A	0,54	H 54 A	0,98	H 53 A	1,13
H 30 A	0,69	H 27 A	1,1	H 37 A	1,21
H 55 A	2	H 48 A	0,43	H 52 A	1,23
H 36 A	1,16	H 29 A	1,21	H 39 A	1,26
H 76 A	0,95	H 72 A	0,9	H 74 A	1,4
H 19 A	0,98	H 23 A	1	H 33 A	1,47
H 18 A	0,69	H 20 A	0,85	H 31 A	1,75
H 60 A	1,1	H 68 A	0,6	H 62 A	1,9
H 64 A	0,95	H 70 A	0,43	H 63 A	2,2
H 77 A	1,1	H 73 A	0,87	H 75 A	2,2
H 58 A	0,9	H 66 A	0,92	H 57 A	2,2

C. Sodium Absorption Ratio (SAR) - 2013

Sample Code	Analysis	Sample Code	Analysis	Sample Code	Analysis
Protected	SAR	Caltivated	SAR	Abandoned	SAR
H77 A	1,3	H 75 A	7,4	H 73 A	13,9
H 76 A	4,8	H 74 A	9	H 72 A	16,2
H 65 A	1,5	H 69 A	9,4	H 71 A	15,8
H 64 A	3,9	H11A	16,2	H 70 A	14,7
H 60 A	1,1	H 63 A	12,5	H 68 A	15,2
H 59 A	12,8	H 61 A	15,2	H 67 A	14
H 58 A	15,7	H 62 A	14,2	H 66 A	14,7
H 56 A	2,8	H 57 A	2,9	H 54 A	14,5
H 55 A	1,3	H 52 A	9	H 48 A	12,3
H 47 A	2,5	H 53 A	5,4	H 40 A	11,5
H44 A	9,7	H 50 A	15,3	H 32 A	14,4
H 42 A	19	H 45 A	9,4	H 46 A	16
H 38 A	3,2	H 49 A	6,1	H 27 A	13,1
H 36 A	16,4	H 41 A	15,7	H 26 A	11,4
H 30 A	3	H 39 A	7	H 29 A	14
H 28 A	9	H 35 A	17,6	H 23 A	11,1
H 21 A	11,6	H 37A	17,5	H 20 A	13,9
H 19 A	14,3	H 34 A	3	H 16 A	14,7
H 18 A	19,7	H 33 A	12,2	H 25 A	13,8
H 14 A	7,4	H 31 A	15,7	H 12 A	14,9
H 13 A	1,6	H 17 A	19	H 43 A	13,6
H 1 A	7,2	H 15 A	4	H 2 A	15

Continue - Sodium Absorption Ratio (SAR) – 2012

Caltivated	SAR	Abandoned	SAR
61	0,93	67	7,97
39	0,95	29	6,1
62	1,11	68	7,84
57	2,5	66	7,94
75	2,9	73	16,2
53	3,2	54	9,65
63	3,2	70	19,1
15	3,87	12	10,4
34	3,9	25	11,86
49	3,98	43	18,12
52	7,97	48	19,31
69	8,1	71	19,6
74	9,1	72	16,64
45	9,8	40	11,45
33	13,2	23	11,05
50	13,9	46	9,78
31	14,5	20	8,64
41	16,8	32	9,92
35	17,1	26	7,9
11	17,3	2	12,9
37	19,3	27	10,1
17	24,8	16	13,5

Appendix (B-III)

Procedures of Soil and water Analysis

A. Water Analyses (Clesceri *et al.*, 1998).

I. pH and EC analyses

Apparatus and Reagents:

- pH and EC – Meter with suitable Electrods
- Buffer solution to calibrate the instruments pH4, pH7 and 1413 µs/cm solutions.
- KCl solusion 3M
- Distilled Water

Procedure

1. pH measurement

1. Calibrate the instrument (pH-meter) by using pH4 and pH7 buffer solution.
2. Rinse the Electrode with distilled water.
3. Read the pH value for the samples directly using pH-meter.

2. EC measurement

1. Calibrate the instrument (EC-meter) by using 1413 µs/cm or any other buffer solutions. But first zeroing EC reading in the air.
2. Rinse the Electrode with distilled water.
3. Read the EC value for the samples

II. Sodium and Potassium analyses

Apparatus and Reagents:

- Flam photometer
- Sodium Chloride (NaCl) dried at 140°C to prepare stock solusion (1000 ppm Na^+) by dissolving 2.543g of NaCl in 1L-distilled water.
- Sodium Chloride (KCl) dried at 140°C to prepare stock solusion (1000 ppm K^+) by dissolving 1.907g of NaCl in 1L-distilled water.
- Distilled Water.

Procedure

1. Prepare different standards from the stock solution according to the estimated amount of sodium/Potassium in the water samples (the linearity for this method is 100ppm).
2. After preparing standards calibrate the Flame photometer and construct the calibration curve
3. Read the value sodium/Potassium
4. Correct the samples reading using the calibration curve to get the concentration.

III. Total Calcium and Magnesium Analysis

Apparatus

Burette, pipettes, conical flask (100 ml) and spoon.

Reagents:

- Standard EDTA solution (0.02 N); Dissolve 3.723g in 1L distilled water (make standardization for this solution by $CaCO_3$)
- Ammonia buffer solution Dissolve 67.5g NH_4Cl in 572 ml NH_4OH_2 and then complete the volume to 1L using distilled water in a volumetric flask.
- Eriochrome black T (EBT) indicator Dissolve 0.5g EBT and 4.5g hydroxylamine hydroxylamine hydrochloride in 100ml high purity ethanol.
- Distilled Water.

Procedure

1. Take 10 ml of sample
2. Add 2ml (40 drops) ammonia buffer solution to keep pH=10±0.1 to prevent precipitating of $CaCO_3$ and Mg $(OH)_2$
3. Add 3-4 drops EBT indicator.
4. Titrate with (0.02 N) EDTA solution until the color change from dark red to clear blue.
5. Record the valume of EDTA used.

Calculations

N1 V1 = N2 V2

Where

N1: Normality of titrant (EDTA).

V1: Volume of titrate (EDTA) used until color changed.

N2: Concentration of total Ca and Mg or Calcium in 10 ml.

V2: Volume of sample (10 ml).

Total Ca + Mg (meq/L) = (N1 V1 = V2)*1000

Total Ca + Mg (mg/L) = Total Ca + Mg (meq/L) * 50

IV. Chloride analysis

Apparatus

Burette, pipettes, conical flask (100 ml) and spoon.

Reagents

- Potassium Chromate (K_2CrO_4) indicator Dissolve 5g K_2CrO_4 in 100ml distilled Water then add $AgNO_3$ 0.01N until a definite red precipitate is formed After 12 hours filter.
- Silver Nitrate ($AgNO_3$) 0.01N Dissolve 2.395g $AgNO_3$ in 1L distilled water store in dark bottle (make standardization for this solution by NaCl).
- Distilled Water.

Procedure

1. Take 10ml of sample
2. Add 3-4 drops K_2CrO_4 indicator.
3. Titrate with (0.01 N) $AgNO_3$ solution until the color change from yellow to gray orange
4. Record the volume of $AgNO_3$ used.

Calculations

N1 V1 = N2 V2

Where

N1: Normality of titrant ($AgNO_3$).

V1: Volume of titrate ($AgNO_3$) used until color changed.

N2: Concentration of chloride in 10 ml.

V2: Volume of sample (10 ml).

Chloride (meq/L) = (N1 V1 = V2)*1000

Chloride (mg/L) = Chloride (meq/L) * 35.45

V. Carbonate (CO_3^{2-}) and Bicarbonate (HCO_3^-) Analyses.

Apparatus

Burette, pipettes, conical flask (100 ml), spoon and pH-meter.

Reagents

- Sandard sulfuric acid (H_2SO_4) 0.02 N carefully 0.6ml H_2SO_4 to 500ml distilled water then complete to 1L (make standardization for this solution by $NaCO_3$).
- Phenolphthalein indicator solution for CO_3^{2-} analysis dissolve 1g phenolphthalein in 100ml pure alcohol.
- Bromocresol green indicator solution for HCO_3^- analysis dissolve 5g of bromocresol green in 100ml pure alcohol.
- Distilled water.

Procedure of Bicarbonate

1. Take 10ml of sample
2. Measure the pH oft he sample
3. If the pH of the sample less than 8.3, the concentration of carbonate in sample equal zero.

Note: the pH of the samples in this research was less than 8.3, and then the concentrations of carbonate in the samples of this research were equal zero.

Procedure of Bicarbonate

1. Take 10ml of sample
2. Add 3-4 Bromocresol green indicator.
3. Titrate with (0.02 N) H_2SO_4 solution until the color change from blue to yellow.
4. Record the volume of H_2SO_4 used.

Calculations

N1 V1 = N2 V2

Where

N1: Normality of titrant (H_2SO_4).

V1: Volume of titrate (H_2SO_4) used until color changed.

N2: Concentration of carbonate or bicarbonate in 10 ml.

V2: Volume of sample (10 ml).

Carbonate (meq/L) = (N1 V1 = V2)*1000

Chloride (mg/L) = Chloride (meq/L) * 35.45

VI. Sulfate Analysis

Apparatus

- Spectrophotometer
- Volumetric flask (100ml)
- Cuvettes Quartz (cell)
- Magnetic stirrer.
- Measuring spoon capacity 0.2 to 0.3.
- Stopwatch (0 – 30 min 12 h. 0.5)

Reagents

- **Buffer soluation:** Dissolve 30g magnesium chloride $MgCl_2.6H_2O$. Sodium acetate $CH_3COO\ Na.3H_2O$. 1g potassium nitrate KNO_3. And 20ml actate acid CH_3COOH (99%) in 500 ml DW and make up to 1L volumetric flask.

Procedure

1. Prepare different standards from the stock solution according to the estimated amount of sulfate in the water samples (the linearity for this method is 100ppm).
2. After preparing standards calibrate the Spectrophotometer and construct the calibration curve
3. Read the value sulfate
4. Correct the samples reading using the calibration curve to get the concentration.

B. Soil Analyses

I. pH Analysis (Bashur and Antaine, 2007).

Apparatus

- pH meter with Combined Electrode
- Glass rod
- Glass Beaker.

Reagents

A. Deionized Water

B. pH 7.0 buffer solution.

C. pH 4.0 buffer Solustion

Procedure

1. Weigh 50 g air-dry soil (<2-mm) into a 100-mL glass beaker.
2. Add 50 mL DI water using a graduated cylinder or 50-mL volumetric flask.
3. Mix well with a glass rod, and allow to stand for 30 minutes.
4. Stir suspension every 10 minutes during this period.
5. After 1 hour, stir the suspension.
6. Put the Combined Electrode in suspension (about 3-cm deep). Take the reading after 30 seconds.
7. Remove the Combined Electrode from the suspension, and rins thoroughly with DI water in a separate beaker, and carefully dry excess water with a tissue.

II. Electrical Conductivity (Bashur and Antaine, 2007).

Apparatus

- Vacuum filtration system.
- Conductivity bridge.

Procedure

1. Prepare a 1:1 (Soil: Water) suspension, as for pH determination.
2. Filter the suspenstion using suction. *First.* Put a round Whatman No. (Suitable) filter paper in Buchner funnel. *Second,* moisten the filter paper with DI water and make sure that it is tightly attached to the bottom of the funnel and that all holes are covered.
3. Start the vacuum pump.
4. Open the suction, and add suspenstion to Buchner funnel.
5. Continue filteration until the soil on Buchner funnel starts cracking.
6. If the filter is not clear, the procedure must be repeated.
7. Transfer the Clear filter into a 50-mL bottle, immerse the Conductivity Cell in the solution, and take the reading.
8. Remove the conductivity cell from the filteration, rinse thoroughly with DI water, and carefully dry excess water with a tissue.

Note

1. Reading are recorded in milli-mhos per centimeter (mmhos/cm) or deci-Siemens per meter (dS/m). The use of unit deci-Siemens is preferred over the unit milli-mhos. Both units are equal, that is, **1 dS/m= 1 mmho/cm.**

2. Reading are usually taken and reported at a standard temperature of 25°C.

3. Check accuracy of the EC meter using a 0.01 N KCl solution, which should give a reading of 1.413 dS/m at 25°C.

III. Organic Matter

A. Main method to measure percent of the organic matter (Walkley, 1947)

Equipment

1. 500mL Erlenmeyer flasks
2. 10mL pipette
3. 10and 20 mL dispensers
4. 50mL burette
5. Analytical balance
6. Magnetic stirrer
7. Incandescent lamp

Reagents

1. H_3PO_4, 85 percent
2. H_2SO_4, concentrated (96 percent)
3. NaF, solid
4. Standard 0.167 M $K_2Cr_2O_7$: Dissolve 49.04 g of dried (105°C) $K_2Cr_2O_7$ in water and dilute to 1 L.
5. 0.5 M Fe^{2+} Solution: Dissolve 196.1 g of $Fe(NH_4)_2(SO_4)•6H_2O$ in 800 mL of water containing 20 mL of concentrated H_2SO_4 and dilute to 1 L. The Fe^{2+} in this solution oxidizes slowly on exposure to air so it must be standardized against the dichromate daily.
6. Ferroin Indicator: Dissolve 3.71 g of o-phenanthroline and 1.74 g of $FeSO_4•7H_2O$ in 250 mL of water.

Procedure

1. Weigh out 0.10 to 2.00 g dried soil (less than 60 mesh) and transfer to a 500 mL Erlenmeyer flask. The sample should contain 10 to 25 mg of organic C (17 to 43 mg OM). For a 1 g sample, this would be 1.2 to 4.3 percent OM. Use up to 2.0 g of sample for light colored soils and 0.1 g for organic soils.
2. Add 10 mL of 0.167 M $K_2Cr_2O_7$ by means of a pipette.
3. Add 20 mL of concentrated H_2SO_4 by means of dispenser and swirl gently to mix. Avoid excessive swirling that would result in organic particles adhering to the sides of the flask out of the solution.
4. Place the flasks on an insulation pad and letstand 30 minutes.
5. Dilute the suspension with about 200 mL of water to provide a clearer suspension for viewing the endpoint.
6. Add 10 mL of 85 percent H_3PO_4, using a suitable dispenser, and 0.2 g of NaF, using the "calibrated spatula" technique. The H_3PO_4 and NaF are added to complex Fe^{3+} which would interfere with the titration endpoint.
7. Add 10 drops of ferroin indicator. The indicator should be added just prior to titration to avoid deactivation by adsorption onto clay surfaces.
8. Titrate with 0.5 M Fe^{2+} to a burgundy endpoint. The color of the solution at the beginning is yellow-orange to dark green, depending on the amount of unreacted $Cr_2O_7^{2-}$ remaining, which shifts to a turbid gray before the endpoint and then changes sharply to a wine red at the endpoint. Use of a magnetic stirrer with an incandescent light makes the endpoint easier to see in the turbid system (fluorescent lighting gives a different endpoint color). Alternatively use a Pt electrode to determine the endpoint after Step 5 above. This will eliminate uncertainty in determining the endpoint by color change. If less than 5 mL of Fe^{2+} solution was required to back-titrate the excess $Cr2O_7^{2-}$ there was insufficient $Cr_2O_7^{2-}$ present, and the analysis should be repeated either by using a smaller sample size or doubling the amount of $K_2Cr_2O_7$ and H_2SO_4.
9. Run a reagent blank following the above procedure without soil. The reagent blank is used to standardize the Fe^{2+} solution daily.

10. Calculate C and organic matter percentages:

A. Percentage easily oxidizable organic C:

% C = [((B-S) * M of Fe^{2+} * 12 * 100))/ grams of soil * 4,000]

B = mL of Fe^{2+} solution used to titrate blank.

S = mL of Fe^{2+} solution used to titrate sample.

12/4,000 = milliequivalent weight of C in grams.

To convert easily oxidizable organic C to total C, divide by 0.77 (or multiply by 1.30) or other experimentally determined correction factor.

B. Percentage organic matter (OM):

% OM = % C/0.58 = % C*1.72

B. Secondary method to validate and assessment (Walkley, 1947)

Apparatus

Magnetic stirrer and teflon-coated magnetic stirring bar.

Glassware and pipettes for dispensing and preparing reagents.

Titration apparatus (burette).

Reagents

A. **Potassium Dichromate Solution ($K_2Cr_2O_7$), 1N**

- Dry *potassium dichromate* in an oven at 105°C for 2 hours, cool in a des- iccator (silica gel), and store in a tightly stoppered bottle.
- Dissolve 49.04 g *potassium dichromate* in DI water, and bring to 1-L vol- ume with DI water.

B. Sulfuric Acid (H_2SO_4), concentrated (98 %, sp. gr. 1.84).

C. Orthophosphoric Acid (H_3PO_4), concentrated.

D. Ferrous Ammonium Sulfate Solution [$(NH_4)_2SO_4.FeSO_4.6H_2O$], 0.5 M

Dissolve 196 g *ferrous ammonium sulfate* in DI water, and transfer to a 1-L volume, add 5 mL concentrated *sulfuric acid*, mix well, and bring to volume with DI water.

E. Diphenylamine Indicator ($(C_6H_5)_2NH$)

Dissolve 1 g *diphenylamine indicator* in 100 mL concentrated *sulfuric acid.*

Procedure

Weigh 1 g air-dry soil (0.15 mm) into a 500-mL beaker.

1. Add 10 mL 1 N *potassium dichromate* solution using a pipette, add 20 mL *concentrated sulfuric acid* using a dispenser, and swirl the beaker to mix the suspension.
2. Allow to stand for 30 minutes.
3. Add about 200 mL DI water, then add 10 mL *concentrated orthophos- phoric acid* using a dispenser, and allow the mixture to cool.
4. Add 10 - 15 drops *diphenylamine indicator*, add a teflon-coated magnetic stirring bar, and place the beaker on a magnetic stirrer.
5. Titrate with 0.5M *ferrous ammonium sulfate* solution, until the color changes from violet-blue to green.
6. Prepare two blanks, containing all reagents but no soil, and treat them in exactly the same way as the soil suspensions.

Calculations

Percentage **Organic Matter** in soil:

1. **$M = (10/V_{blank})$**

2. **% Oxidizable Organic Carbon (w/w) = $[((V_{blank} - V_{sample}) * 0.3 * M) / W_t]$**

3. **% Total Organic Carbon(w/w) = 1.334 × % Oxidizable Organic Carbon**

4. **% Organic Matter (w/w) = 1.724 × % Total Organic Carbon**

Where: M = Molarity of ferrous ammonium sulfate solution (approx. 0.5 M).

V_{blank} = Volume of ferrous ammonium sulfate solution required to titrate the blank (mL)

V_{sample} = Volume of ferrous ammonium sulfate ferrous ammonium sulfate solution required to titrate the sample (mL)

Wt =Weight of air-dry soil (g)

0.3 = 3 × 10-3 × 100, where 3 is the equivalent weight of C.

Note

1. For soils high in organic matter (1% Oxidizable Organic Carbon or more), more than 10 mL potassium dichromate is needed.
2. The factors 1.334 and 1.724 used to calculate TOC and OM are approximate; they may vary with soil depth and between soils.
3. Soils containing large quantities of chloride (Cl-), manganese (Mn--) and fer- rous (Fe++) ions will give higher results. The chloride interference can be eliminated by adding silver sulfate (Ag2SO4) to the oxidizing reagent. No known procedure is available to compensate for the other interferences.
4. The presence of CaCO3 up to 50% causes no interferences.
5. To assessment the results of Percentage Organic Matter in soil was used others Procedure like ashing and bu using the same above procedure but different materials.

IV. Saturated Paste Soil (Bashur and Antaine, 2007).

In order to determine of soil analysis should be make paste soil to take the leachate to make analyses.

Apparatus

1. Sensitive balance, beaker 600 mL, spatula, graduated cylinder 100 mL.

Procedure

1. Taken a sample weight of 300 g of soil air-dry in Beaker 600 ml.
2. Add distilled water using a graduated cylinder gradually and mix it by spatula with the soil until it reaches the saturation and record the volume of water which was added.
3. Know the saturation point through surface gloss of the paste and the reflection of light on it and gonorrhea slow when move the beaker and slip the paste from the spatula without that cleave.
4. Leave the paste after the prepared with covered for the next day to water absorbing by soil
5. The next day is to make sure the dough she was still in the case of saturation if it observed that its stiff must add a little water, to be placed in the filter
6. Filter the suspenstion using suction. *First.* Put a round Whatman No. (Suitable) filter paper in Buchner funnel. *Second,* moisten the filter paper with DI water and make sure that it is tightly attached to the bottom of the funnel and that all holes are covered.
7. Start the vacuum pump.
8. Open the suction, and add suspenstion to Buchner funnel.
9. Continue filteration until the soil on Buchner funnel starts cracking.

Note:

1. Water extracted from the paste od soil was used to measure: EC, PH, Trace and heavy elements.
2. Is treated with it like a water sample but without dilution just in case of non-appearance of read or result are mitigated taking into account the mitigation.